Science or Society?

Steve Sturdy

SCIENCE STUDIES UNIT
UNI OF EDINBURGH

Crucible: Science in Society
General Editor: Robert M. Young

SCIENCE OR SOCIETY?
The Politics of the Work of Scientists

Mike Hales

Design by Tony Fry

Pan Books   London and Sydney
in conjunction with
Channel Four Television Company Limited

First published 1982 by Pan Books Ltd,
Cavaye Place, London SW10 9PG
© Mike Hales 1982
ISBN 0 330 2692 3
Printed in Great Britain by
Richard Clay (The Chaucer Press) Ltd,
Bungay, Suffolk

CONTENTS

Supplementary Material

Acknowledgements

These people have been involved directly in the making of this book, either in discussing ideas in the early stages or as authors of articles which influenced the way I dealt with certain topics. They all helped and I'm happy to thank them: Gavin Browning, Simon Schaffer, Karl Figlio, Ludi Jordanova, Lyn Allison, Les Levidow, Edward Yoxen, Maureen McNeil, Kate Hinton, Brian Easlea, John Krige, Tim Lang, Graz Baran. Thanks also to Antonia Ineson for picture research and permissions and special thanks to Margot Waddell and Les Levidow for last-minute support. Cynthia Cockburn led a superb seminar which I was able to listen to on tape, and this gave me my 'line' on computers in The Print (chapter 8). Bernard Leach had a key influence in the way I set out to write the book and Tony Fry, as designer, has been deeply involved in making it work. Ken Green put ideas into my head late on, which I hope I have been able to use to make the book more useful, and Bob Young, as editor, has worked very closely with me all along.

Thanks should also go to the Crucible unit at Central Independent Television, which comprises Jane Cousins, Nick Davidson, Lawrence Moore, Liz Nash, Lila Petrou, Denis Postle, Morag Roberts, Sandra Sedgbeer, Julie Sheppard, Terry Walbe and Bob Young.

Acknowledgements to the following for permission to reproduce quotations: the estate of the late Sonia Brownell Orwell, Martin Secker and Warburg Ltd. and Harcourt Brace Jovanovitch Inc. for 'James Burnham and the Managerial Revolution' by George Orwell, Methuen and Co. Ltd for *Contrary Imaginations: A Psychological Study of the English Schoolboy* by Liam Hudson, Ward Lock Educational Co. Ltd for *Writing and Learning Across the Curriculum 11–16* by Nancy Martin and others, The Union of Physically Impaired Against Segregation for their *Policy Statement*, Farrar, Straus and Giroux Inc. and Souvenir Press Ltd for *Listen, Little Man* by Wilhelm Reich, Routledge and Kegan Paul Ltd for *Ideology and Utopia* by Karl Mannheim and *Inside the Primary Classroom* by M. Galton, B. Simon and P. Croll, Hart Davis McGibbon and Co. for *In the American Grain* by William Carlos Williams, and Penguin Books Ltd for *Understanding Children Writing* by Carol Burgess and others and *Science and The Nation* by the Association of Scientific Workers.

Acknowledgements to the following for permission to reproduce illustrations and photographs: BBC Hulton Picture Library (page 158), Farrar, Straus and Giroux Inc. and Souvenir Press Ltd (page 129), Archives of the California Institute of Technology (page 86), Janis Goodman and the Campaign Against Depo-Provera (page 162), *The Guardian* and Peter Clarke (page 13), the Royal Society (page 70).

Using This Book

This book examines the relationship between expert knowledge and society. Its chapters, which range widely over issues in the social relations of science, technology and medicine, consider ways in which this relationship can be transformed. Although no chapter goes very deeply into specialised arguments, each one provides a basis for further discussion, analysis and study. As a guide to going beyond the boundaries of the book, there are some aids at the back: a 'Users' Guide', a short bibliography, a list of useful addresses and an index.

Chapters – and sometimes sections within a chapter – can be taken on their own, as short essays on specific topics. Nevertheless their sequence is significant and the argument is developed and followed through the prologue up to the final chapter. Later chapters are based on the preceding discussions and examples, and make more use of the specialised terminology that is needed to address the issues. When specialised terms are used in a chapter they are often more connected with the chapter's place in the sequence than with the topic being discussed, so that it should be possible for readers to re-read earlier chapters using ideas from later ones to explore the topics in another way.

If later chapters seem to stand less well on their own because of specialised terms, then the Index should help in exploring where those terms come from and what they mean.

There are four principal lines of argument in the book and most chapters sit mainly within one or another of these. (The divisions are spelled out in the Users' Guide.) The book assumes no specialist background in science beyond secondary school. What it does assume is a personal experience of the contradictory and often oppressive nature of state and corporate institutions: office work, industry, the civil service, welfare agencies and at the very least, school.

Tomorrow Has Been Cancelled

THINGS MOVE FAST THESE DAYS, most of the time faster than we can watch. What is technically possible or about to become possible is often beyond imagining, unless you happen to be a worker in that field of development or a science-fiction writer. When the possibilities are presented to us, as on TV in programmes like *Tomorrow's World*, it tends to be in a Gee-Whiz fashion: 'Just look what we're doing for you now!' Outsiders, including people who may be experts in some other field, may try to keep up and at the same time be critical, but then the tendency is to slide into an exactly opposite mood: 'O My God, look what they're doing now!' In this book an attempt is made to steer a different course, and especially to focus not so much on the *things* – the hardware – but more on the reasons and the social connections. This is after all what the best science-fiction writers do.

ROBOTS AND SATELLITES

'The chip' has its wonders constantly extolled by the media: the circuits of a computer compressed on to a thin sliver of silicon a few millimetres square, 64,000 units on a single chip today, over a million expected by 1990; by the end of the century, memory chips and computing chips at virtually zero cost – that is, *money* cost. What the cultural cost may be is another issue. As work proceeds on the chip itself and on ways of making it cheaper to use, which means making it easier to programme in order to perform any of a vast variety of functions, applications also proceed apace. We already know them in some recognisable forms: TV games, space-invader machines, home computers. Your new washing machine may contain microchips that will supplant and, manufacturers hope, make obsolete the electro-mechanical control gear in older models. They will provide new options at the touch of a button, and can be serviced by replacing a whole circuit board in one go. Designers of these machines take care to make them 'user friendly', that is, to make their uses *obvious*. What happens to those who use them is less obvious. Take the situation of children using microcomputers in schools.

First the bad news. It is possible for some children (maybe one in five gets keen enough to give up spare time) to become very involved with computing. It can be weird. One sixteen-year-old stopped talking to people, headed straight for his home terminal after school and could barely be persuaded to take meals. To stop him at the end of the day they had to cut off the mains power to the whole house. That is not normal, but just how many of our households are 'normal'? And just how could new subcultures like that of computer freaks aggravate or challenge existing patterns of domestic life and social relationships? The cultural interactions may be abnormal, or they may be 'normal' but regrettable – like the way that one fourteen-year-old played truant from school and used computing know-how to rip off commercially marketed programmes to accumulate a private library of pirated software. At the last count, that pirate library had about 1,000 items, current market value around $50,000. Before security was tightened up, a commerical stock market data service was accessible by phone, for free, if you knew how, and this bright star did. The new technology here is just a new gambit in the old game of private enterprise.

And the good news? A black pupil, reading at a level four years behind his chronological age, found mastery and respect through the microcomputer. He is the school's best programmer. He teaches younger kids the ropes, and as he says, 'I love these machines. I've got all this power at my fingertips. Without them I don't know what I'd be. With them I'm somebody.' A fifteen-year-old who often teaches teachers sees it this way too. 'It's a sort of mutual doorway. The barriers between adult and child are broken, and it's person to person. Nobody's looking down on anyone, they're looking each other right in the eye.' Chippewa Indians use a computer to help children learn the ancient tribal language. Deaf children have found new autonomy through the visual display/keyboard combination.

Clearly, it all depends. Teaching with computers can be just a new way of putting children through hoops, performing drills, and at this level the computer is just a technical fix whose days (like all technical fixes) are numbered. Then again, computers can be used to present kids with 'simulations' of quite complex situations, such as fighting a forest fire, or navigating across an unexplored ocean.

3

On one hand, this can be useful 'off-line' learning about potential real-life situations. On the other, it can be an avenue into unreality, like a lot of existing TV programming. Some computer-graphics experts project the fantasy of complete Feelorama computer systems in the home within twenty years: wire yourself up and fly the frozen mountains of Neptune, explore the seabed, gun down an enemy, be tortured to death by the Inquisition, then spool it back and play it again.

What matters most though, is whether users are consumers only, or have the direct and indirect power to make the machine perform for them in response to their own needs. It could happen that the present generation of school-age computer freaks is having a ball, soaking up all the flexibility in the technology, only to leave the same old barren landscape for the next wave: routines, fixed uses, software fully developed and locked-in. It could also happen that new elites are in the making. A computer executive says proudly, 'If you were born before 1965, you're going to be out of it.' A sixteen-year-old sees it that way too, 'Either you get the hang of it or you don't.' This is not about the machines at all but about culture, and power, and how shifts in the one produce and repro- duce shifts in the other. It's not up to the machines whether micro- chips give more power to the people, nor should it be up to the designers and entrepreneurs. It's our business. We can't afford to be mystified by the hardware trappings and the jargon; it's all about social relations. We *can* understand that, once we learn to see things that way.

'Robots and satellites' is the title of this section. These are the areas where microchip technology is likely to hit us first as adults, unless we have children who are already coming home from school full of things we don't understand. Well established tendencies in industrial production tend to shift work in two opposed directions – towards 'dial watching' and towards humping and packaging. The modern chemicals factory is a good illustration, with white-overalled control room operators in high-grade jobs on the one hand, and loaders and packers in low-grade jobs on the other, servicing the inputs and outputs of the manufacturing process. Automation through low-cost electronics – the chip – will be seized upon by manufacturers to take this further and into new sectors. But at the same time, the low-grade work is in for a thrashing. Routine jobs requiring precision have been targetted for robotisation: you may have seen film clips of robot welders on the Fiat car assembly line, or robot sprayers painting a chair in fifteen seconds flat. Packing and materials handling, where women do many of the jobs, are under the hammer too. The high-grade jobs are less seriously im- plicated, but as more powerful cheap computing comes on the market, 'skilled' jobs will be made more idiot-proof (and that makes you an idiot, right?). Just in case, automatic logging and surveillance

4

of movements and actions is a facility built into computer-control systems, such as 'electronic mail' systems which route communications through a computer network.

With electronic mail we move from the sphere of robots (automation of physical tasks) to the sphere of 'satellites'. Satellites may sound a long way from where you are, but that's the whole point. New technologies for processing information (speech, written words, coded data, visual images, all reduced to a common digitalised electronic form) can take the scale of the systems actually involved right out of the sight of any individuals or groups of users. Out of sight is out of mind until we are confronted by the consequences, such as new work routines, because the computer 'needs' data fed in a particular way, or a new job definition which has become necessary because the computer is being used to achieve greater efficiency and the employee is to be relieved of 'waiting time', or unemployment because more computers mean fewer typists, mail clerks, designers, typesetters, machine operators, unskilled workers, skilled workers, and so on.

Where do satellites come into all this? They are just transmitting and receiving stations orbiting in space, conveniently placed to relay information without using expensive or inefficient means of surface communication. Satellites are a glamorous sounding part of an expanding nation- and world-wide network of electronic information media, including television channels, and of course computers. Globally, the network gives new power to big users of data such as bankers, multinational firms, governments, military institutions and the police. They can capture and keep more information on more subjects, and they can rework it and retrieve it faster.

Back on the ground, more jobs are set to become 'computer aided': computer-aided learning was mentioned earlier, and we saw how conditional that 'aid' can be. In manufacturing firms the stages of designing a product, making working drawings, scheduling production and actual control of the machine-tool on the shop floor can all be linked together and eventually telescoped into one stage by using computer-aided design (CAD) and computer-aided management (CAM). On the shop floor, with or without these white-collar rationalisations, there may be machines controlled numerically by computers (CNC, computer numerical control) and groups of machines organised for rapid and easy programming and reprogramming of flows of work material (FMS, flexible manufacturing systems). The new elite coming up through CAL (computer-aided learning) may feel quite at home in tomorrow's world. But they're not the ones whose jobs and skills and working lives are being disintegrated now under the flow of new techniques and technologies.

Don't think you're safe in the office. Warehouse stock control and fetching is now fully automatable. You may have seen automated retail stores like the Argos chain in England. Banks are working hard at increasing their interest from your money and keeping their own labour costs down (i.e., fewer jobs) by computerising financial transactions. Do you have a cash-till card? There's more where that came from. Banks can resume Saturday opening because computers are making a smaller staff go further.

You'll notice, I hope, that most of the time this discussion is about the content of work, rather than machines. That's the way I mean to go on. But in general terms the prospect of robots and satellites can be put quite starkly. Any human activity which can be translated into a series of specifiable operations – a set of strict rules – can in principle be programmed and given to a computer-controlled machine. The big question then is, who makes the rules, and whose game are we in?

BUGS AND SPARES

As far as 'bugs' are concerned, the prospects here too can be put quite compactly. Almost any chemical substance, whether drug, synthetic fibre, fertiliser, animal feedstock, or fuel, might in principle be produced by engineering a bug to do the work as part of its own 'natural' metabolism. The technique behind this is genetic manipulation: taking the DNA (deoxyribonucleic acid) of a microscopic organism – the material which carries the code for physical reproduction – snipping it chemically, and inserting some different genetic material. The result is a modified organism that does slightly different things, 'naturally', and the resulting genetic material is called recombinant DNA, or rDNA. rDNA is a key growth point for chemical manufactures in industries based around molecular genetics, biotechnology.

Research overheads are high, since all this is new work, so firms are heading in two broad directions, towards low-tonnage, high-value products, and towards bulk products. Drugs come in the first category – interferon was one of the first products of rDNA biotechnology. Fertilisers, animal feedstocks and fuels come in the second. Apart from the price characteristics of these markets there are in profit terms two things going for biotechnology. One is that the conversion of materials by bugs is efficient because it proceeds via the internal energy-conversion processes of living cells, and consequently biotechnology can be an energy-saving (money-saving, profit-enhancing) alternative to conventional synthetic routes. The second factor is that bugs can often be bred which feed off cheap materials, materials in world surplus such as grains.

But control is a much underrated issue. The modified bugs used in rDNA work are supposed to be able to survive only under controlled conditions, so that if any escaped they would die instead of proliferating in the wild. Technically, however, the controls are far from watertight (bacteria-tight?) and there is constant strong pressure from interested corporations (which include the world's largest chemicals multinationals as well as newer specialist outfits) for further slackening of legal safeguards and watchdog procedures because they reduce the pace of research. But what would the consequences be if, for instance, a copper-eating bug (under development for a potential biotechnological mining process) escaped into the urban sewage system for a feast of copper pipe? Could it be killed without killing other bugs on which we normally depend? Could it be confined? Could the disaster be pinned on anybody? Could it be spotted in time?

Control is crucial at another level too. rDNA (recombinant DNA) techniques applied to plants rather than micro-organisms have produced 'revolutions' in farming over the last thirty-five years, mainly in the form of higher-yielding varieties of grain. The power of these techniques – enhanced by cloning, which speeds up the rate at which variations can be produced – now makes possible the wholesale customising of nature at a price which big companies can afford. Seeing the opportunities, chemicals multinationals have been moving in on agricultural suppliers and seed-breeders, buying them up. As feedstock for the chemical giants' future bug-based manufacturing processes, and as foodstuff for people and animals, the seedstock of the world is moving gradually, but faster, under the control of a small number of companies. In a sense this is 'only' more of what we can see looking back over the past hundred or so years of industrial growth. But the geographical scale of the new connections and the intimate level at which they operate in the biological economy makes this an urgent political challenge.

Problems with 'spares', that is, with advanced techniques of human body engineering such as 'spare part surgery' and test-tube fertilisation, are much more widely recognised though they tend to be seen primarily as moral rather than as political issues. At the level of public debate matters seem to be very confused. Cloning is 'bad' (making copies that are biological identical) because it threatens cultural identity, yet the major focus of development in cloning is on speeding up the production of new varieties of trees and plants – which is 'good' because it offers technical fixes for food shortages, pollution and deforestation. Test-tube fertilisation is 'good' because it may enable the childless to have children, yet this too is manipulation of a biological form – in some ways like cloning – and political considerations are fearsomely complex. On the day when the first test-tube twins in Britain were born, the director of the Centre of Law, Medicine and Ethics at London University told family planning doctors that facilities for test-tube birth should be denied to some people. The new technique opens up a new level of potential discrimination, and a new sphere of struggles between the power of experts and ordinary people.

Horror stories are easy to dream up in the sphere of bugs and spares. What if a market is opened up by sharp operators in spare parts? (This could happen *now*.) The poor and desperate in the USA, as well as in much poorer countries, make part of their living by selling their blood to blood banks. Are commercial spare-part banks out of the question? Kidneys and corneas of living people are already on sale in Brazil. And is physical violation of the identity of the poor then ruled out? What if rich white women decide that they want to buy the use of wombs of impoverished black women from Third-World countries, to have babies which they are too busy or squeamish to carry? (This could happen tomorrow.) Violence to human identity on the most intimate scale is partly what the politics of bugs and spares is about. Horror stories are easy to dream up – but nightmares are not easy to legislate away. After all, the medical industry in the USA was worth $180,000,000,000 in 1978, and markets have a way of opening up under that kind of pressure. How will the new biotechnologies be incorporated into the balance of power between professionals in medicine and research, companies in chemicals, drugs and construction, and the consumers – you and me?

WINDMILLS AND NODULES

The other sector of technical development where things are moving fast is energy and materials – windmills and nodules. 'Windmills' stands for the whole range of energy-generating technologies which now exist at some level of development, from conventional coal/oil/electricity through established nuclear fission and Great White Hope fusion, to 'alternative' sources in wind, sun, waves, geothermal energy, biogas (methane from cowdung), biomass (green, growing things) and so on. Despite the starvation of funds (in 1980–81 all the 'renewables' research in Britian cost £11 m., compared with nuclear power's £170 m.), it seems likely that energy from renewable sources could match that from nuclear reactors within twenty years and could cost much less. As this picture began to emerge over the past decade, shifts have taken place in the energy business so that the big oil giants are no longer *oil* giants: they own *energy* resources and patents across the whole range, ready to move in any direction which begins to show profit advantages.

'Windmills' means more than power sources, for energy conservation is a key aspect of strategy. For use in homes ten watts can be conserved at the same cost as installing new capacity to produce one watt. Home energy conservation is not unproblematic, however. Millions of Americans are presently confronted with evidence that a vapour released by the foam cavity insulation in their houses may cause cancer. Money costs are inseparable from social and health 'costs'. Energy efficiency in domestic and industrial processes can be lifted by tighter control of chemical reactions – this is what microprocessor-based ignition control systems in cars do. Bugs can bring greater energy efficiency to industrial manufacturing processes. Clearly there is a great deal of scope for innovation. To research-based multinationals that means scope for profit and that, in turn, means that the political trends we noted with bugs are present also in windmills. More power in fewer hands, worldwide, mediated through professionals committed to more sophistication through technique. This is not a situation that existing democratic institutions are equipped to cope with.

'Nodules' means materials which have to be extracted from the places where they occur in the earth's crust or on its surface. Mineral reserves which are too costly to extract at present may be brought into the extractable range, and therefore made into *resources*, through new developments in mining technology such as robots, or processes which turn coal into gas for easy handling. New resources may be recognised because of developments in conversion technologies, for example, trees (already a threatened resource on the world scale) can become chemical feedstocks in place of oil, thanks to new options in bug-based manufacture. Mineral reserves in inhospitable terrains may become exploitable. Nodules of metals (containing manganese especially) exist on the ocean floor, so that strong reasons exist for mining companies to push robot remote-handling techniques in this kind of application.

There's a snag which you might not necessarily anticipate. Ask yourself: Which interests are powerful and wealthy enough to drive exploration into the most unnatural human habitats – the deepest oceans, airless space? Answer: the military, in a endless drive for mastery over territory. The oceans are, of course, the places where nuclear submarines lurk, pawns of terror, and materials handling and surveying in the ocean deeps is a business of great interest to the Pentagon and other military centres of control. Minerals we may need, and manganese nodules we may have to find if we want manganese at the right price. But part of the price, and it doesn't appear in the market, is the yet closer convergence of military and industrial interests. More research, more development? Inevitably. More political threats and challenges? Inescapably.

Things move fast these days, often faster than we can see. What makes the whole situation today so fraught is that we are also having to learn to see new connections; cultural and political *relations* in the 'things' which are the technological tip of the iceberg. We not only have to keep moving in a literal sense, as jobs shift from sector to sector, country to country, with unemployment as the shameful staging point. We also have to move mentally, to cope with more details and directions of change, and many quite new national and local and international relationships between people, firms, governments and nature. There is a desperate need for new voices with which to speak about what is happening, and what we wish to happen. Inseparable from this, there is a practical need for new

9

places to look from and exchange ideas in and to act in, so that wishes actually begin to touch what moves 'out there'.

It is early days yet, and the voices and the places are only just coming to be recognisable. Yet it is also very late. Tomorrow's world is an extension of today and today was set in train yesterday. By the time that we see the latest gadget on television's *Tomorrow's World* it is too late, it's already happened, new gadgets are in the pipeline and they too will be out there confronting us before we have come to terms with them. Tomorrow has been cancelled by the massive projection of today's social relations and trends into a massive apparatus of scientific research and development. WE DO NOT NEED MORE OF TODAY; we need a tomorrow, a new day. This book is part of the search for voices and places that we must make our own, so that tomorrow can be our own too.

Robots — Satellites (r.h. pair) Bugs — Spares (centre pair) Nodules — Windmills (l.h. pair) Apocalypse

1 Mankind's Little Helper?

SCIENCE IS SO CENTRAL in modern life that it is hard to know where or how to place it. It is a constant source of private worries – terrified by the Bomb, confused by arguments over nuclear power, dazzled and dazed by the multiform appearances of microchip technology. When we feel we understand it at all we are awed by the prospects of manipulated life through birth technology and spare-part surgery, and mystified by the power now held over things called genes which no one has ever seen. These issues can seem too distant to grasp, unless we happen to be a worker displaced by a robot or an infertile woman or a physically impaired person, when the issue comes unpleasantly close but is not necessarily more intelligible.

At other times we confront facts of modern life, hardly aware that there is any 'issue' at all. We enjoy our jet flights to Spain, take

---

HEADPIECE Maria's robot 'double' in the film *Metropolis* (1926). Rotwang, the male scientist-magician proclaims, 'I have created a machine in the image of man, that never tires or makes a mistake ... Now we have no further use for living workers.' Maria mediates between workers in an underground city where they toil in indescribable conditions to service the surface metropolis. Covered with flesh, the gleaming warrior-robot Maria, made by Rotwang, incites workers to violence, to destroy the underground workplace city and so themselves. The workers burn her at the stake and the real Maria escapes; but to what purpose ... ?

pleasure in our new colour telly, rely on our domestic 'labour saving' machines, worry about health, medicines, diet and sex, love and hate our work, join in or resent trade union action over 'rationalisation', and get hung up over dealings with doctors, schools and government officials. Many times, consciously or not, we appeal to 'science' to come up with solutions to our problems, without seeing that 'science' lies within them already and without trying to come to terms with that contradiction. Science, technology and medicine (in this book 'science' often stands for all three) are not a clear-cut success story.

Yet the list of advances is long, and few of us would feel entirely easy about turning our backs on them and similar ones in the future: antibiotics, agricultural techniques, transport systems, computers, electric light, and the bodies of knowledge which lie behind these and other facts of industrialised life. By and large we value these things; but we don't feel entirely comfortable.

'Gee Whiz' programmes on the TV, our teachers in school and the papers we read tell us one thing, while our experience often suggests another. Where we work, new machines and methods are designed, developed and introduced with a fanfare of science and scientific management which threatens identities, skills, traditions and jobs by the hundred thousand without offering us new prospects. Old and new, workplaces (including the places where most women work most – homes) bristle with hazards which can and do maim and kill, and we get told it's our fault if they catch us out. On the street where we live it can be no better, children having to

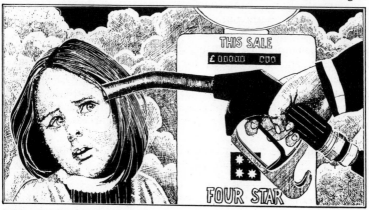

(Drawing by Peter Clarke, *Guardian*, 1 October 1981)

breathe in lead from petrol and who-knows-what else from the stack of the nearest factory. Nuclear and toxic wastes trundle down our shopping streets and past our back gardens. Safe at home, how much of the food we eat is really good for us – and how can we know? Many of the things we use, from washing machines to con-

13

traceptives, cars to computer 'aids', can seem some or most of the time to be part of a trap.

The things, activities and ideas that are connected with science – and it's difficult to think of many that are not connected in some way – are somehow mysteriously distant even though we live our lives right in the middle of them. They won't easily offer up their secrets to a curious intelligence which wants to see the wood as well as bump into the trees. And this is a serious difficulty, because it is not a viable, short-term option to abandon the things, the activities and the ideas. They exist and reproduce themselves day by day, in our work and our leisure and our thoughts. We depend on them and even if we want to end up with something different they are our existing cultural stock and we have to start with them. We're in a fix and Scotty, up there on Starship Enterprise, will *not* beam us up. We cannot look to the technicans – or the enterprises, for that matter – for answers. They are parts of the problem.

What makes the problem so hard to grapple with is the contradictoriness of our attitudes, perceptions and responses, and this lies in the contradictoriness of our experience itself. At a personal level the institutions and products of scientific research, scientific medicine, scientific design, scientific management and scientific opinion can make life painful and hard, even at the same time as particular institutions and products ease our pain and labour in particular circumstances. The drug we are taking to shift that hacking cough doesn't make us feel any easier about the 'permitted antioxidants and colorants' in our margarine; the WD-40 which we use to cold-start the car in the morning ('spin-off' from the US space programme) doesn't reassure us about the spy satellite which sits in its geostationary orbit over our heads; and no amount of 'labour saving' technology at work will make us feel OK if we are women and 'scientific' arguments say we should go home and give the kids maternal attention so that men can have the jobs, or if we are black and we hear people 'proving' that we should not be doing the job anyway because whites are genetically better equipped and have higher IQs.

Our experiences of science are contradictory because science is contradictory, not a single 'thing' but a multiplicity of practices which connect with each other and our lives in a multiplicity of ways. It's better to speak, a lot of the time, about scienc*es* (and technolog*ies*, and medical practic*es*) so that the detail and the differentiation is emphasised. This book sets out to map some of this tangled web of practices, and some of the alternative activities and ideas which exist or might be made to exist.

TWO TRAINS

How to think about science? At a general level there is a difficulty we have to struggle with, in terms of the images which our culture

14

offers us for picturing science to ourselves. For example, there's a gruesome account of how nomads, members of 'untouchable' castes, in India fell foul of modern life. Wandering in a hot deserted region through which there runs a single railway line, the Indians would often pillow their heads on the cool steel rail to sleep during the hottest part of the day. Trains were infrequent but a number of people were killed through this innocent practice. That image might, for many, sum up modern technology. On one hand the onrushing force of the machine, unable to stop in time even when

'You come round the bend, you know it's the end . . . the fireman screems and the engine just gleems . . .' Words by Robert Hunter to 'Casey Jones', from the Grateful Dead album *Workingman's Dead*. (Photograph: The Twentieth Century Limited, New York Central Railway)

danger is perceived, innocent victims paying the penalty. On the other hand the problem for the railway administration: How can we educate these people for their own safety? There is a parallel with many situations closer to urban lives as in industrial health and safety for instance. But the image is not particularly powerful or clear, and an alternative will show why.

In David Lean's film *Doctor Zhivago* there is a scene in which people are travelling across a barren and deserted steppe, crowded into a long train of wagons. They are shunted into a siding off the single track and wait, inexplicably. From the horizon comes a faint sound, growing with a blurred shape, closer and louder and faster, until it becomes the blunt, speeding, vibrant iron-clad Red train of the commissar. The close beating of its cylinders possesses your heart, a huge star on the smokebox leaps from the screen. Faster than it approached, it is gone, and the space it occupied is more silent and empty than before.

As an image of technology there is more truth in this one than in the other, for this is not an image of *technology*. The Red train is manifestly the vehicle of power, and the relationship between the people, shunted into a siding, and the machine is manifestly a political one. This is the case with all technology and with all artefacts and structures of culture, including languages and theories. The great power of the film director's image in this case lies not only in the way that it evokes the breathlessness and fear of being close to a big machine, but also in the fact that the machine's social meaning is explicitly present. The ability to see things this way is not easy to develop or to live with, but it's something we all need, despite the fact that most of us are not film directors. All the world's a stage, and everything in it seems to have been produced as a prop for some action. The problem, most of the time, is getting into the action.

Where does this leave the other, profoundly unheroic, situation – the untouchables who died sleeping under the wheels of unexpected trains? As posed, the solution to the problem for the railway's administrators appeared quite obviously in the form of 'education'. Education is, classically and essentially, an administrator's solution to problems (the vicious irony here being that the Latin, *educatio*, means 'to bring out' whereas administrators understand education as putting *in*, installing certain skills, traits, inhibitions and information). What other solutions might there be? Over hundreds of miles of desert it's impractical to install and maintain adequate fencing. And you couldn't close down the railway!

WHY NOT?

Why not? Well, that would be totally unrealistic.

On the contrary, 'Why not close down the railway?' is the only realistic question, because only this kind of question forces the answer to address the interests that are embedded in the situation.

16

Why not close down the line? Well, there is the economic power that the line gives to remote townships and regions, and the personal power that it represents and enhances for railway magnates and politicians, and the institutional power that it carries for state functionaries, and the direct benefit it endows (and the power it imposes) on many mostly poor Indians whom it puts within markets and other new 'communities'.

How can all these interests be counterbalanced by the mashed brains of a few untouchables? That is not a question which it makes sense to try to answer here. It is not, in terms of 'balance' or simple reason, an answerable question. For one thing the interests themselves are not coherent or 'balanced' internally, and for another, the 'solving' of problems like this is a matter of negotiating on the ground with the groups of people involved, not sitting down with a pencil and paper. The point, as far as this book is concerned, is to notice how the 'why not' question can open up insights into the real historical intractability of problems connected superficially at least with technology.

Where Does Nature Start?
As a strategy for probing science this is quite generally appropriate. It may seem obvious enough in relation to technology because machines and hardware in general are made, that is, made to some purpose, and the 'why not' question brings out the often hidden purposes. But it works with science and medicine too because – and this is a very difficult but fundamental understanding to establish – nothing that we live with is 'given'. It is all made. Not just the machine-world of the city, the motorway, the global village of the airwaves, the ideal home, the hospital, the factory, the arsenal; the preponderance of machines makes the manufactured nature of these institutions obvious. But disease too, the kind which is called iatrogenic disease, is made. The term means 'disease caused by the system of medical treatment' and includes physical damage caused by doctors in their attempts to cure people by surgery or chemical therapy. The definition of iatrogenesis could also be extended to include the addictive dependence of people on the 'care' of experts and expensive institutions, and the degradation of individuals' sense of responsibility and actual or permitted power to deal with their own health and ill-health. Beyond that, our concepts of disease and our diagnostic classifications are made, in the sense that medicine is a system of ideas, a way of seeing living nature in distress. There are systems of medicine other than our own, in other cultures.

Similarly, although it is a trite and sexist expression, the environment is man-made. Not only in the sense of being constituted largely by artefacts which are arranged in certain patterns, but also in the more fundamental sense that the immediate environment of any individual or social group is partly made up of its own waste or pollution.

The world we live in and which humans, as humans, have always lived in is not simply 'given', as a thing beyond culture, which is imposed on it. But the non-givenness of nature is in fact even more profound than this. Nature itself is made. This is the sticking point for many people in thinking about science, but we must get past this point if we are to explore its deepest roots. It does not mean that humans, rather than God or the Big Bang, started the whole thing rolling and created the Earth, the Sea and All That In Them Is. There surely is a natural world out there, and in a radical and humbling sense it has nothing at all to do with us. But that is not really the important point because in those aspects where it has no connection with us and our lives, we can know nothing of it anyway. Neither science nor any other part of culture (except mysticism) has anything to do with that. Humans are creatures who attach meaning to the world and therefore the world of science – nature and society both – is constructed in terms of what it means for us.

Take a thunderstorm. We have not made the thing, nor can we unmake it. But it is not just a thing. It can be a threat, or an awe-inspiring experience, or an interesting meteorological phenomenon, or a challenge to Faustian instincts – a wildness to be brought under control. This latter, characteristically, is the voice of much of modern science. But whatever kind of interest sciences have in nature, the world they approach is always a distinctly human nature, a world given shape by meanings and purposes. The conclusion then is vital: to understand sciences and technologies and medicine, we must understand people's needs and purposes and identities. If we hope to understand science we must aim to understand society.

WHY KNOTS

The 'why not' strategy in exploring science is a bit like the five-year-old's perpetual 'Why?' but with at least one difference. The effect (and the purpose?) of the five-year-old's tactic is to push you back and back until you're up against the wall and then you say 'Because I say so' or something equally authoritarian and final. In contrast, an effect of the 'why not' strategy is that it leaves behind at each stage of the pursuit a residue of potential alternatives. In retrospect these might be sifted and gathered together to define a potential alternative practice to that being criticised.

Let's look at the washing machine. Someone has set out to do a load of washing and the machine breaks down. What happens? It could possibly be mended, but probably not. Why not? For one thing, because it's not designed to be fiddled with. The larger mechanical components may be accessible (with a lot of humping and lifting which only some people could do) but electronic components, and the 'best' washers nowadays are microcomputer controlled, are a closed secret. In any case it may be necessary to travel to the other side of town or another town to buy spares. The fact is, that the machine and its associated servicing system are not designed for amateur repairs. It is a commodity, and part of the power that the manufacturer has wielded is the power to commit you further to purchasing commodities when it goes wrong. You can't repair it from the box of straightened nails and old hinges that Grandad used to keep in the cellar (not that nails are re-usable these days, they're softer and they break). You have to call out the service engineer – and pay for the privilege of having him walk across your threshold.

But let's say that the machine is mendable. If the user is a woman, she still probably won't mend it. Why not? Because it's not women's work. From her youngest days right through her schooling and her adult contacts with other women and men, she has been steered away from an involvement with working on machines (as opposed to buying and using them). She has not been encouraged to study technical subjects, she has been laughed at or patronised if she showed an inclination to go in for 'boyish' interests. She feels she knows nothing about how to go about the task, so she curses and fumes and waits for the engineer and humps the wet load to the launderette.

She doesn't leave it for her husband to fix – why not? Well, she tried that before, with the washer, the iron, the hoover. He said, 'Can't it wait till the weekend, I've been out at work all day?' It took two weeks to badger him into trying it. And eventually she realised that he was embarrassed by the assumption that *he* could fix it. As Diane Harpwood asks in *Tea and Tranquilisers* (Virago, 1981), how is it that a man can run a factory, design a space rocket or split the atom and still come home and fail to be able to mend the washing machine? The answer is that despite the place of machines in male culture, 'practical' interests are discriminated against in school. So-called higher education is even more removed from the problems faced by individuals in everyday life. Schooling, while it reproduces social discrimination against girls, helps men to come to terms with everyday life little more than it does women.

Why have a washing machine in the first place? You can't do without a washing machine. Why not? Granny used to manage, without present-day 'easy-care' fabrics. But her granddaughter has to work part-time outside the home, to bring more money in. And granddaughter lives in a society where expectations of 'cleanliness' – cleanliness as manifested in regular lines full of washing rather than a once-a-year clean-out – have grossly inflated since Grandmother's day, thanks to the efforts of domestic science experts and manufacturers of washing machines, washing powders and easy-care fabrics. In any case it's more convenient, especially without a car, than trailing to the launderette if there is one. And anyway, a shiny new washer makes the kitchen look more 'modern', and that seems to matter.

Behind the 'why nots' there is a trail of unfulfilled alternatives to the washing machine as a boon, a debt, the location of a chore, a self-awarded sign of status, a charge (when faulty) and a stage for playing out the complex relationships between those who do domestic labour, those who go out to work, those who make money out of selling things, those who work for them, those who make clothes dirty and those who look on. Why not have a public service to replace the individual burden in money and time and frustration? Why not make it a collection service so that dependency is not shifted simply on to another commodity, the car? Why not redirect general education so that women's future competence is not discriminated against, and so that general everyday needs are explored and supported in preference to the specialised and local needs of professions and 'higher' education? None of the questions has an answer. But the movement from part-answer to further question yields a mapping of the meanings and values and relationships embedded in a piece of machinery, a thing.

The same kind of inquiry could be carried out into industrial health and safety. Why not make industrial processes safe without question, instead of boxing-up workers in masks and earmuffs, discriminating against them in screening, burdening them with responsibility for undue care and attention, and making them fight for compensation when the inevitable happens? It would be possible to 'why not' military research and development, or selection through IQ tests, or the national health care system, or schooling. Even basic research in astronomy, which uses sophisticated computer-imaging techniques derived from military programmes. Why not? If any activity or thing or idea has a value then it has a connection with the rest of society. And if it has a connection, that connection could be different. Why not?

THE LONG REVOLUTION

Underneath – and often on the surface – science is at the centre of many urgent public issues as well as policy issues, concerning nuclear weapons and nuclear power, new technology, technical training, medical care. It is also at the centre of deep philosophical issues of objectivity, realism, fact and value. This book concentrates more on private worries than public issues, and so proposes that 'philosophical' problems can and should be resolved within the context of everyday life and work in an industrialised society, rather than letting them be reduced to merely academic matters. Such a resolution will require a practical understanding of the relationships between experts and laypeople, working and schooling, assessments of progress and senses of tradition, and active choices in the face of the intractability of nature and society. Through practical connections such as these the philosophical and the policy issues link up with each other, and with private worries.

Private worries settle at many levels and on many grounds, such as being in ill-health, being worried sick by what heads of government might do to us with their sabre-rattling and their nuclear arsenals, being out of work or living with a person out of work, right down to not being able to do the washing because the machine's broken and it costs £15 to get the engineer in. But perhaps most deep rooted, because it is a component of all these other worries, is the feeling of inadequacy because you're not sure what you really know, and whether what you know actually counts or means very much. In some ways, in a highly complex society with contradictory internal processes, the 'Don't Knows' may be the most profoundly honest sector of social opinion. Yet it is a disgrace not to be in the know.

In this unhappy position – feeling ignorant and ashamed of it – it is all too easy to defer to 'the experts' who exude an image of knowing. This is what one teacher felt about approaching the controversial and clouded issues of science in society:

> To teach science and society materials so that they inspire involvement, even social criticism, can be unfamiliar to the point of being frightening. [To the teacher, at least.] In science we are used to accepted, consensual theories; neither Ohm's Law nor cellular respiration can be matters of fervent individual opinion.

Today, maybe not. But in their origination?

> How convenient it would be if only our scientific notables could expound the 'correct view' on the social problems of science in the same straightforward way!

Now, this may be some kind of embarrassed joke, but even so it indicates a deeply mystified understanding of the status of scientific knowledge in relation to issues of moral and political rights and wrongs. It is the narrowness of experts which qualifies them to pronounce on specific technical topics and this gives them no special status as commentators on society and history. If anything it is a disqualification. In any case, the experience of the past century should surely be enough for anyone to see that the experts cannot be trusted to define significant issues on our behalf.

What are the implications when science is invoked as a saviour or solver of problems? Can you oppose, for instance, nuclear weapons and still appeal to scientific method or technological advance when you hit an obstacle at home or at work? There is no point in pretending that this is a simple question with a known answer, for the social relations of science are profoundly contradictory. Science produces or supports:

> innovation, elites, pollution, deskilling, decisions, Faustian confidence, explosion of knowledge, force and aggression, inequality

21

and theoretical insight.

And by the same token, science selectively and systematically produces or supports:

traditions, mass living, wonder-substances, technological progress, risk, ignorance, violation of persons, fine operational discrimination and ideology.

It is clear that science is not a take-it-or-leave-it matter, for it is clearly not a single 'thing'. It is also clear that, although there may be nest-feathering, graft, empire-building, incompetence and self-consciously cynical manipulation in science, confusion is not essentially a matter of personal motivations of scientific and technical workers. To understand the systemic relationships, we have to look at the system.

When we invoke 'science', what do we mean?

There is no *science*. We have to be clear which practices we are referring to, which *sciences* (or technologies, or medical practices).

Do we mean sciences now, or then, or in the future? In the UK or the USA or the USSR or India or somewhere else?

Do we mean, not a particular practice but *abstract knowledge* as a cultural form? Or expert-power, which goes with abstract knowledge but is not the same thing?

Do we mean science as *science-and-technology*, that is, as a 'high' technology force of production?

Do we want to invoke the idea of an ombudsman, or a public assaying service which is neutral, available to all, guaranteed by procedures and law?

Or do we just mean *rigour*? If so, shouldn't we be more precise – rigorous – because there are as many ways of being rigorous as there are arts and disciplines, and more.

Which images, models, values, definitions, practices do we mean to build on in the future?

This point is crucial. If life and work are being recomposed continually through science and technology, and this is manifestly the case, then science and technology themselves are also undergoing continual transformation. If what 'science' is is far from fixed, are the *values* constant? And in any case, do we want them? The cultural challenge is how do we take up and develop the positive values that 'science' has symbolised (objectivity, productivity, power, progress, intellectuality, security through understanding) and how do we damp out the tendencies in social practice which bring criticism to science, such as remoteness and detachment, waste, domination, risk and dangerousness as an ethic of exploration, unfeelingness? Understood practically, this means how do we transform existing sciences and technologies, and how do we create new ones? Science, as it exists and as an ideal, is just raw material for this transformation. For the transformation itself we must look, not to science but to the people – their needs, intentions and identities. On the basis of those we should pick and choose our knowledges.

The idea of science as a 'long revolution' is one that is taken from a book of the same name by Raymond Williams. He describes it as a threefold process consisting of an industrial revolution, a cultural revolution, and a democratic revolution. To see sciences' past as an industrial revolution is a commonplace, and with all the new technologies that are bubbling under at present (robots and satellites, bugs and spares, windmills and nodules) it is clear that the revolution is continuing. What has been said so far, about the importance of understanding needs, purposes and values in planning for science in the future, underlines the aspect of cultural revolution, so that the political aspect of science – the democratic revolution – may be the least obvious. This book is about all three aspects of the long revolution in science, but especially about this latter. It is about power, and how power is entrenched in science.

The scientific aspect is obvious in some politicial struggles. Military and riot control technology (nerve gases and teargas, government surveillance computers, electronic battlefields, 'smart' missiles) are wheeled in to 'fix' conflicts. The superpowers seem willing to fix us in an even more horrifying way through the escalation of the nuclear 'balance'. On an everyday but still horrifying scale, experts fix workers and blacks and women with technical arguments 'proving' that the workplace is safe and the new process necessary for 'efficiency', or that women should be in the home or using 'digital dexterity' in clerical jobs, or that blacks are genetically and necessarily different and 'therefore' inferior. These vicious and deadly serious games go on and on. Men are innately aggressive, hierarchies of power and authority are natural . . .

These are roles of science in the ideological and manifestly coercive reproduction of a social order and its inequalities. The main focus of this book is on political processes of a different kind; it is mainly about power over work and processes of production more generally. It is about school, office, factory and farm, and about home too, as the place where production is normally seen in a deprecating way as 'reproduction', a secondary function. The exercise of power is not a one-way process, power is always being renegotiated, and the experts' technical fix is only one gambit just as Luddism is only one response. (Luddism is absolute refusal of an oppressive 'progress' embedded in new workplace technology; in history 'General Ludd' was a machine-breaker.) One of the main aims of this book is to show that work is also a sphere of reproduction, a place where forms of power are reproduced and struggled over. The emphasis is twofold: on science as central in work, and on science itself as work. Sciences are part of society, and political interests and struggles move and develop through science just as much as through any other central sector of social life and work.

AT THE SECOND INTERNATIONAL CONGRESS of the History of Science, held in London in the summer of 1931, one of the eight Soviet delegates planned to speak on 'Electrification as the Basis of Technical Reconstruction in the Soviet Union'. His paper was ruled out of order because it dealt with the future rather than the past, and historians were unaccustomed to accepting the future as part of history. Some historians at least have learned better since then, but such an approach is still unusual enough to warrant an explanation. As a book essentially about the future of science, this one has a lot in it about the past. It does not contain plans and blueprints and drafts. It explores values and images and relationships, and takes the past as its material. Reworking such materials is a task for history – the members of a culture as a whole – and only a very crass author, or a planner, would attempt it in the abstract. This book is an exploration of material for thoughts and actions around science in the 1980s.

2   Science and the People (1)
Science 'On the Grain Front'

LOOKING ACROSS HISTORY and across societies it is possible to see many different relationships between scientific knowledge and social action, scientific workers and other members of society. We need to learn to think positively and creatively about evaluating different modes of science and selecting directions for the advancement of knowledge, since all directions cannot be followed and all do not

HEADPIECE In proletarian costume, T. D. Lysenko surveys a field of grain. Tatlin's tower, a monument to the Third International, was never in fact built. Made of 'iron, glass and revolution' it would have been a hundred metres taller than the Eiffel tower, and equipped to project movies and messages on to clouds.

25

have the same values. In this chapter and the next, we shall plunge straight in and survey two examples of scientific development aimed consciously, at one level anyway, at the needs of the people. In both cases the countries are what we often call 'underdeveloped', but their underdevelopment has developed in different ways. Both have a degree of industrialisation. The first case, in this chapter, is 'Lysenkoism' in Soviet Russian agriculture during the 1930s and 40s. The second case, in chapter 3, is 'the green revolution' in wheat and rice growing in India during the 1960s and 70s. Taken together they give us some ground for discussions of the relationships of science, experts, culture and politics. The argument is quite dense and textured, because science and technology are embedded in the particulars of real history, with as many factors involved in shaping science as there are in shaping society. Indeed, the same forces shape both. Unless science is considered in this fully historical way, it will continue to mystify by its claims to be above the actual conflicts in society.

Genetics is an area of research which has made enormous strides during our lifetimes, and it lies at the centre of both situations. Genetics is the branch of science which studies the ways in which inherited traits in living organisms are controlled and specifies the 'building blocks' of the body which make possible the 'phenotype', the actual characteristic behaviour which finally appears in the mature organism. Over the last thirty years the science has moved into specifying the sequences of building blocks in DNA (deoxyribonucleic acid) the genetic material. The basic material is called the 'genotype', and for the last forty years in Western science it has been accepted that the genotype determines the phenotype, and the genes set limits to what an adult organism looks like and can do. Anything that the adult organism 'learns' cannot be passed on to its offspring genetically. Changes in the phenotype leave the genotype unaltered. In humans, for example, sperm and eggs do not change as a result of what we *learn* during life. If they did we would not pass on the same characteristics at, say, age fifteen as at age thirty. In terms of scientific knowledge, this is the principle challenged by Lysenkoism.

When Francis Crick and James Watson, who shared a Nobel Prize for their DNA work in 1962, published the now famous 1953 paper outlining the 'double helix' structure of DNA, there was no journal in the USSR in which it could be reprinted. There was no field of genetics research. No genetics texts were published between 1938 and the early 1960s, so that whole generations of biologists and doctors were educated without knowledge of the current state of research into the fundamental organisation of living matter. Since the 1930s, when there had been Soviet geneticists of international reputation, some cultural disaster had taken place, destroying a whole sector of work and links with an international research community.

The Grain Front

In the late 1920s Russia under Stalin was faced with a terrific agricultural crisis. A massive drive for industrialisation meant that there was a growing urban proletariat which needed feeding by the efforts of a diminishing rural peasantry. But also the USSR needed foreign exchange to buy the machinery with which to create an industrial base. What could be exported? Grain, if there could be a surplus. The political decision was made that there *would be* a surplus, that agricultural science must create it, and that the industrialisation programme would go through.

As a residue from Tzarist days there was an intelligentsia within the very lopsided social structure of Soviet Russia, and among this stratum there were biological scientists, university based and European oriented. The Soviet leader, Lenin, had early made a decision to compromise with experts such as these from pre-revolutionary days, even though many of them had strongly opposed the Bolshevik revolution. It was a practical necessity that unsympathetic managers and technicians should continue to be employed, if the economy were not to break down completely However, the members of the biological establishment were unwilling to accept the dramatic practical task set by political decisions of the time. State planners' response to this was, in effect, to build an alternative establishment, to freeze and squeeze and take out the scientific elite and establish a new order of experts, more committed to the practical needs of the country.

The alternative was built around T. D. Lysenko, a 'horseback plant breeder', son of a peasant from the Ukraine and trained at the Uman School of Horticulture and the Kiev Agricultural Institute, who worked at the Ukranian Institute of Selection and Genetics in Odessa around 1930 on a technique known as vernalisation. Winter varieties of wheat normally fail to bear ears and ripen when sown in spring. Vernalisation was a way of treating sprouted grains at near-freezing temperatures so that winter varieties could be effectively sown as spring crops. Thus it was hoped to increase the effective length of the agricultural year and increase potential production. Lysenko's conclusions began to be controversial (though his practical techniques had already been criticised) during the mid-30s because they were in direct conflict with the prevailing views on heritability of characteristics. Orthodox genetics maintained – and still maintains – that characteristics 'learned' in the phenotype are not passed on through the genotype. But Lysenko claimed that the characteristics of vernalised wheat (resistance to cold and 'unnatural' sowing seasons) were transmitted to future generations 'under the right conditions' which were never clearly specified. This was just what the planners needed to hear.

The polemic between Lysenkoists and orthodox geneticists, who were among the residual intelligentsia, was bitter. Lysenko was 'a muzhik's son', 'illiterate' and 'ungrammatical' – not that these were legitimate scientific considerations according to the orthodox ideology of science; but then, the limits of 'science' are precisely what the controversy is all about. Legend has it that one of his papers began, 'Take a bucket . . .' From the Lysenkoist side the members of the genetics establishment looked like 'priests' of 'aristocratic and lily-fingered' science. Lysenko was ignorant, blustering and opportunist; but he was also devoted to finding a solution to the problem of feeding the proletariat and raising foreign exchange. Against the

T. D. Lysenko in the ascendant. (Photograph: frontispiece from Julian Huxley, *Soviet Genetics and World Science: Lysenko and the Meaning of Heredity*, Chatto and Windus, 1949)

detached, elite, abstract, cultivated values of the old middle class, Lysenko represented the engaged, enthusiastic, practical, self-taught values of the proletarian image. Lysenkoism was not what it is often called – an 'affair'. It was part of a full-blooded attempt at cultural revolution.

In his speech 'On the Grain Front' in 1928, Stalin declared against the assumption that 'even the greatest of all revolutions, the October revolution, cannot change economic forms overnight'. As this line was more widely imposed it fell to Lysenko to become director of his institute and his voice became more powerful. A prominent critic of high international standing, N. I. Vavilov, President of the Lenin Agricultural Academy in the early 30s, was arrested in 1940 and died as a result of his imprisonment. Lysenko acceded to higher positions and numerous honours until, in 1948, the Presidium of the USSR Academy of Science passed a resolution virtually outlawing any biological work not done within a Lysenkoist framework. The Presidium was not uninfluenced by a message from Stalin which arrived at the correct theatrical moment, expressing his own personal support for and interest in the policies and theories of Lysenko. Critics recanted, were removed from positions, or were arrested, and through the apparatchiks' (bureaucrats') machinery Lysenkoist positions came to dominate not only plant breeding but also livestock breeding, animal husbandry, forestry and – through genetics – biology and medicine generally. When Crick and Watson's paper was eventually slipped into the Soviet literature in an obscure journal in nucleotide chemistry it immediately sold out; but for at least a decade after Lysenko's assumption of power in 1948 there was an officially policed amnesia which related to all matters within one of the most spectacularly successful areas of modern science.

CULTURAL DISASTER

A cultural disaster certainly, especially for a State which attaches so much importance to scientific and technical progress. But how are we to locate this extreme form of development within the framework of scientific practice and policy more generally? Lysenkoism was one phenomenon of a massively authoritarian and bureaucratic form of State power, and conditions in all cultural fields under Stalinism were extreme in the extent to which a top-downwards line was imposed, and in the institutional and physical violence used to police that line. Nevertheless we need to understand what there was in science and society which made it possible, and eventually necessary, for a whole way of looking at the world to be removed from the cultural map. First, however, a couple of qualifying comments concerning the extent of Lysenkoism's deviance from established norms.

Firstly it is important to recognise that basic issues in genetics were unresolved at least until the 1940s. The founding work of the modern era, Julian Huxley's *Evolution: The Modern Synthesis*, was not published until 1942 and the Darwinian mechanism of evolution by natural selection was regarded as discredited by the author of a European standard history of biology in 1928, the year of Stalin's 'Grain Front' speech. The mechanism of Darwinian evolution is again under question in the 1980s. We must not, therefore, harshly judge the scope of intellectual options open to Russian agriculturalists in the 1930s and 40s by simply projecting backwards a set of norms which became established in Western Science well after the main lines of disputation had been drawn in the USSR. In addition, relations between scientists in Russia and elsewhere had become politically overdetermined by the mid-1940s, as the power of the Soviet state began to be taken seriously as a political factor by Western politicans and militarists. The Cold War following World War II served to entrench the isolated position of Soviet genetics built up during the 1940s.

A second qualifying point is that for all its theoretical crudity and questionable technique, Lysenkoism does not seem, on the basis of evidence available to us, to have done much major damage to levels of output in agriculture. The techniques of Lysenkoist agriculture were wasteful and inefficient, such as sowing several seeds in one hole, so that only the strongest one would grow. They could not deliver the goods promised on the grain front. But neither do they seem to have diminished wheat yields. The ineffectiveness of Lysenkoist agriculture eventually came home to roost in 1964 when Nikita Kruschev was dismissed for reasons which included the stagnation of agricultural productivity. But before that time the economic case seems to have been too uncertain to undermine the political determinants and the institutional entrenchment of Lysenkoism's empty promises in agricultural policy.

The Intelligentsia

Class divisions in Soviet society are central in weighing up the kind of developments that could reasonably have been expected in agriculture during the 1930s and 40s. Within society generally divisions were highly polarised by the desperateness of the economic situation. Most scientists were accustomed to a cosmopolitan lifestyle established before the revolution, and while its excesses had been curtailed by the Soviet state they still retained an identity as an intellectual elite. They moved in separate circles from workers and Communists, looking outside Russia for cultural recognition, as the intelligentsia had done in Tzarist days. They were also institutionally detached from the running of an economy which would not support them in the style to which they aspired.

Divisions also existed within science, polarised around the division between theory and practice. Geneticists in academic establish-

30

ments were able to develop sophisticated analytical approaches to problems of producing higher-yielding or more robust crops through selective breeding. But such conventional scientific approaches could not deliver the goods fast enough since they necessarily had to operate over a number of (plant) generations. In any case, the scientific elite were unwilling to compromise their methods of research – and their lifestyle – in submission to the urgent need for faster practical advances. It thus passed to scientists who were committed to the Revolution, and to Party careerists, to work at whatever pace events demanded, and these workers were outsiders in terms of the intellectual establishment of genetics. Thus a general class division within society, between revolutionaries and bourgeois experts, reproduced itself within the science of genetics as a radical split between theory and practice. The theoreticians would not engage in the required practice. The practically-minded were intellectually unsophisticated as far as genetic theory was concerned. A rift which might have been bridged and filled in under other circumstances was widened by the political forces which had to be activated if things were to move fast enough for Russia's industrial 'take off'.

Proletarian Experts

The political situation was in fact even more complex. One conclusion drawn by political leaders about the struggle on the grain front was that peasants were witholding output. Rather than forwarding it, as earmarked by planners, for urban consumption and export, peasants were eating and storing it themselves as indeed they would have done in any time of surplus, because that was how peasant culture worked. The situation was aggravated, however, first by requisitioning and then by forced collectivisation, so that by the 1930s the largest sector of the population – the peasants – were deeply suspicious of the motives of the Soviet state and entrenched in opposition to the dictates of planners, unable to take into account (through blindness or urgency) the traditional ways of working and thinking. In particular the kulaks – larger, richer farmers – strongly resisted and resented the inroads of the State into their privileges of consumption and authority exercised through land ownership and private wealth.

T. D. Lysenko was of course himself of peasant stock, and of peasant sympathies in opposition to the middle classes who were prepared to see the Revolution fail and the peasants continue uneducated, underfed and downtrodden. His approach to agricultural matters was 'unsophisticated', and this might have been a real advantage in making results and policies accessible to common people. But between even the alternative practices of Lysenkoist science and the mass of common people there was a deep, alternative divide of expertise. Stalin had a slogan, 'Technique decides everything', which embodied for him and his supporters in the Revolution an image of the role of science in social development. Research would determine the best solutions to problems, the Party would see that they were implemented, the People would welcome progress. The slogan later became even more telling: 'Cadres decide everything'. Science passes over, in practice, into a matter of elites, so that Lysenkoism was not the supplanting of a bourgeois science by a science of the people, but the replacement of a bourgeois elite of detached experts by an elite with different class origins and connections and a differently validated mode of expertise – political rather than technical.

'Bourgeois science' or 'proletarian science' were both separated from the people *as science*, conceived as the knowledge of experts, generated within an elite culture, out of daily reach of the experiences and views of non-specialists. This separation, implicit in the Stalinist (and bourgeois) notion of science, was powerfully policed and built upon by the dramatic polarisation of classes during the period of Lysenkoism.

'Science' has more to answer for than this, however. We have seen that there was a rift within the scientific 'community' on class lines, widening during the 1930s. We have seen the rift between the Party, with its assumption of expertise, and the peasants. But there was a further dimension which we might see as a rift between society and nature. This was intrinsic in 'progressive' ideas of what science was, but before outlining it we should note what progressivism stood for, in general and in the struggle around genetics. Above all, progressives stood for an optimistic view of human nature and

---

THE LYSENKOIST MILLENIUM

AN UNPRECEDENTED WAVE of enthusiasm swept through the ranks of our agrobiologists, soil scientists, agronomists, zootechnicians, academicians and advanced kolkhozniks. The great festival that had been spoken about at the session [the 1948 three-day enlarged session of the Presidium] seemed to have arrived . . . Never had life made such demands upon the scientist, and never had the demands of life so enthused the scientist. A whirlwind of gigantic activity, of noble tasks, blew into the recently closed and stuffy rooms where yesterday the Mendelists [orthodox geneticists] had engaged in their pettyfogging scholastics. The fields are waiting! The livestock farms are demanding! – this became the supreme law.

---

human society, in the face of fatalistic notions. Reactionary biologists argued (following Malthus, for example) that the poor must stay poor because they are the children of the poor and therefore inherit their presumed biological, inheritable, weakness. Such reactionary views were associated with the scientific positions of bourgeois geneticists throughout the world, and thus with the Western-looking intelligentsia inherited by the Revolution. This was one powerful reason why supporters of the Revolution fought against the theory of inherited characteristics and welcomed Lysenko's 'discovery' that new behaviour could be passed on to the future generations. Connected with this deep opposition to fatalistic thinking, progressives stood for economic change and struggle in the face of arguments which asserted that the USSR was not modern enough to industrialise. Proletarian experts were committed to the idea that through experiment and fearless implementation of plans, radical changes in social order could be accomplished against all the restraints of tradition.

Diamat

Dialectical materialism – 'Diamat', as it became known when established as an official philosophy of the Soviet state – carried the progressive idea of science. The name and the didactic development of the idea came from Friedrich Engels in the late nineteenth century, and were taken up in the USSR under Stalin in the form of a system of 'laws' of dialectical development. These asserted, for example, that changes in quantity eventually burst the boundaries of their form and so pass over into changes in quality – a new order. The key aspect of Diamat was that this scheme of analysis was taken to apply, not just to historical events – such as the Revolution itself as the qualitative change resulting from capitalist accumulation – but also to processes in the natural world. Physics, for example, was taken as a field of knowledge which eventually could be reformulated in terms of the principles of dialectical materialism. Not only could methods of investigation in the natural sciences be read off from Diamat, but also the general form that correct results must take. What it amounted to was an assertion that the laws of all development had been discovered and thus that society (symbolised as abstract knowledge) was henceforward prior to nature. Nature held no more secrets, in principle. Nor did society. A set of texts existed which grasped the 'laws of motion' of society and nature, needing only to be decoded and popularised through the Party in specific contexts: Marxism had attained the status of an applied science.

As a conception of science this pompous, complacent and centralist doctrine is appalling. Yet in its own time it was little more than a displacement into politics of the pompous, complacent and centralist self-image of classical physics at the end of the nineteenth century, when physicists believed that the framework of physical knowledge was essentially complete and held together all other knowledge within a hierarchy of abstract disciplines. Whatever its roots and consequences in European physics (the quantum revolution destroyed it in simple form) this attitude in official Marxism meant that Lysenkoism, as the agency of change in the sphere of biological nature, and the Party, as the agency of change in social nature, were endowed with the 'natural' authority of correct knowledge. If science was the search for unchanging formulations of

33

unchanging relationships, then official Marxism, being scientific rather than merely 'utopian', had to possess the laws of motion of history. On the other hand, if an elite was to transform the running of a whole nation, then powerful techniques for 'action at a distance' were essential. In these respects science (as it was fetishised at the turn of the century) and Stalinism were made for each other.

Culture

Lysenkoism nevertheless failed to grasp the nature of the characteristics of wheat. Entrenched philosophically within a scientistic Marxism, Lysenkoism was narrow in the range of concepts and methods that it could accept as explanations and approaches to nature. Being entrenched culturally in opposition to a bourgeois scientific community narrowed its options in relating theory to practice and theory to existing theory. Political entrenchment in the authoritarian organisation of the Stalinist state closed off options in relationships between experts and other people in society. Just as Stalinism presumed that culture – the established cultures of a cosmpolitan scientific elite or a massive peasant class – could be thrown over and replaced piecemeal by other forms articulated through a minority, so Lysenkoism presumed that nature could be made over into a desired form by an effort of will and application of a material apparatus. Any entrenched culture has an established relationship between knowledge and practical success, though it may not be theoretical knowledge that matters, and it may not be technical – as opposed to political – success which counts in that culture. Therefore, what no science and no politics can afford to discount is the material force and the limited but material necessity of established culture inside or outside science, even when it is clear that things could be different.

Culture is not just 'ideas'. It is also relationships, entrenched and reproduced day by day in practice, between people and people and things (including Nature). And sciences, as parts of culture, must therefore be studied not just as 'bodies of knowledge' which change under some internal dynamic. The political conditions of Lysenkoism were dramatically visible, and harshly punitive for many of the participants in the action. But they were not in any significant sense *outside* conditions, which intruded into a scientific 'debate' which could or should have been left alone. Scientists, both bourgeois and proletarian, were members of class formations and they used their science – their knowledge, privilege, access to facilities – to advance their interests as they saw them, whether that meant resisting an authoritarian and philistine regime or advancing the condition of an ignorant and resistant population. Politics was *in* science, and scientists acted out political roles in the publishing of their papers, the reviewing of their peers' work (as when the cosmopolitans sneered at Lysenko's style) and the selection of their

34

research priorities and intellectual models. At the same time science was in politics, both as a vital force of economic production and as a source of metaphors and models for approaching practical issues. And again, scientists carried it there, both in lobbying for patronage and financial support and in projecting ideas from specialist contexts (such as biological evolution) into wider debates concerning history and culture. This was so forty years ago, in the USSR. It is so today, here.

There is no question of separating politics and science, in any real practical sense. The question is whether we want sciences' relations within society as a whole to be those of Lysenkoism or those of modern Western science or some other way. Why not?

It should be added that there is another motive for dealing early in this book with Lysenkoism. For the past thirty-odd years Western scientists who have not wanted to look closely into the social, political and cultural roots and roles of science have used the scandal of Lysenkoism to scare people: 'Look what happens when politics get mixed up with science'. One purpose in this chapter has been to try to make sense of Lysenkoism in its historical context and to suggest that the kinds of forces which gave rise to it aren't all that different from those at work in other societies and at other times. By the way, Western genetics are the norm in the USSR today, and they are trying to catch up in biology, while having to import masses of grain from America, Canada and Argentina.

3 Science and the People (2)
The 'Green Revolution'

WITH A BLOSSOM AS BIG AS A PERSON'S HEAD and a stink like rotten flesh, one of the world's rarest plants lives in a remote area of the Malaysian jungle. It got into the news because it was seen in bloom by botanists who returned from an expedition there in 1981. In December 1968, forty-two peasants in Tanjore, the heart of India's most prosperous rice-growing area, were locked up in a barn after a fight with other peasants, and burned to death. The two events, which seem to be unconnected except by the macabre, signal the extent and complexity of the issues raised by the Green Revolution.

The juxtaposition may be surreal, but the connection is as follows. The wealthier landowners in Tanjore were reaping the first benefits of ADT-127, a strain of 'miracle rice', and local landless or dispos-

---

HEADPIECE The Rockefeller clan, left to right: 'Junior', David, Nelson, Winthrop, Laurance, John D. III. The tower is a fractionating column, which separates heavier from lighter components in oil – a basic operation in oil refining, the source of the Rockefellers' wealth and power of patronage.

36

sessed peasants demanded higher wages at harvest time as their only way of taking some of the benefit to themselves. As it was the expense of growing the new miracle variety had caused some of them to lose their own landholdings to the moneylenders and large farmers. The landowners refused and imported peasants from another area to work at the old rates. It was some of these who died, and the peasants who fired the barn were locals whose employers had brought them in. Throughout the 1970s the Green Revolution, based on high-yielding varieties (HYVs) of wheat, rice and other grains, has spread to involve many Third World countries and industrialised countries (notably the USA). A growing commercial interest in seed banks of local and wild varieties of grain and deforestation of land (for cultivation under high-yield methods or for domestic and industrial timber supply) have combined during that period to emphasise the importance at a world level of what are called *Vavilov centres* of genetic diversity. (They are named after N.

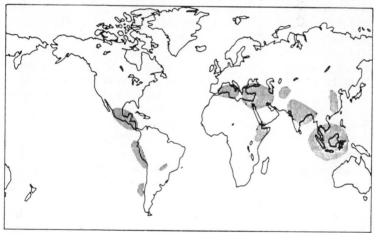

The Vavilov centres of genetic diversity

I. Vavilov, the geneticist who became a victim of the struggle on the Russian grain front in the 1940s.) The jungles of Malaysia and Java constitute one of these Vavilov centres, where parts of the forest approximate to the primeval variety out of which present day species evolved, and the explorers who found the stinking flower were there to investigate the contents of these natural laboratories and storehouses. Thus the surreal connection. More prosaically, we can talk of it in terms of the role of seeds in the world economy today.

THE ROCKEFELLER REVOLUTION
The story starts in Mexico in 1943–44, where four American scientists had been sent by the Rockefeller Foundation to find ways of boosting wheat output. The Rockefellers had had their holdings in

Mexico nationalised, and were perhaps seeking to improve their corporate image and claw back some influence. It has been suggested that for individual scientists who became involved in the research programmes in later years there may also have been another dimension, a kind of guilt or striving for moral credits through helping the poor to feed themselves, to offset the obscenity of the atomic bombing of Hiroshima and Nagasaki. If this were so then there is a deep irony in the fact that the breakthrough in producing new varieties was accomplished by taking germ plasm (crudely, seeds) from Colombia, Australia, the US and Japan and incorporating into it a gene isolated by Japanese scientists.

The gene produced dwarf characteristics in wheat. Strains of wheat which traditionally survive in the tropics are tall and thin. But when heavily fertilised (to produce a higher yield) the head of grain becomes too heavy, the stems bend over forming 'lodges' and the grain won't ripen. The new Rockefeller-financed dwarf varieties were short strawed so lodging was not a problem and large amounts of fertiliser could be used. Indeed, they were essential. Technically, therefore, the problem was solved. Mexican farming eventually switched almost entirely to HYVs of wheat, with the consequence that over a quarter of a century output of wheat has almost trebled and Mexico is now a wheat exporting country. What Lysenko failed to achieve by setting aside modern genetics, the Rockefeller researchers had achieved. The leader of the Rockefeller research, Dr Norman Borlaug, was awarded a Nobel Prize in 1970, two years after the Green Revolution, based on HYVs, had provided the conditions for the riot at Tanjore. It is a real revolution, not just a revolution in technique, because it carries its own political and economic precursors of conflict and transformation.

Unproletarian Science
How the Green Revolution is political may be difficult to see at first. There are no visible conflicts such as were at the heart of Lysenkoist genetics, no controversy about 'bourgeois' or 'proletarian' science. The fight on the grain front in the West was tackled, not by rejecting the sophisticated science of elite institutions but by employing all the resources of orthodox genetics that capitalist conscience money could buy. (The Ford and Kellogg Foundations later joined in.) The science of the Green Revolution is not maverick science, with no past and an unpredictable future, but an extrapolation of an orthodoxy, and it works under defined conditions. Technically, the Green Revolution can deliver the goods. India began to adopt the new genetic package in the late 1960s. In 1979, it was being recommended that the government should start planning for regular grain exports so that stable markets for the surplus could be established. The picture is not entirely clear, but it seems likely that India may become a net exporter of wheat and other

HYVs including rice, and in 1979 there was a record government held reserve of grain totalling twenty-six million tonnes, equivalent to more than half the import of food grains into the Third World four years before. It should be noted, however, that an official surplus, and grain exports, do not mean that Indian people have all the grain they need.

Just a Few Problems

In what sense is the Green Revolution going sour; how does this prime example of science 'for the relief of man's estate' generate messy and violent politics? We could say that the scientists are naive. Less charitably, we might say that experts and those who manage them are mealy-mouthed. James Grant, President of the Overseas Development Council, Washington, stated in 1979 that because yields per acre in India were one third or one quarter the level in more advanced countries this meant that India 'could be producing at least three times more grain per acre . . . if it could overcome certain problems of organisation, finance, and applied research'. The language is that of the technician, the obstacles are 'problems'. This mentality is also that of another commentator, from the UN Food and Agriculture Organisation, who talked of India's 'obsolete social structures' – as if social structures were products of a manufacturing plant, designed, engineered, and narrowly defined so that one generation of products can be marketed as a replacement for a previous generation. This technician-speak can turn easily into mealy-mouthed apologetics for the technicians' own cultural modes and existing social structures – the pattern of land ownership, for example. This way of seeing things certainly does not help researchers to get to grips with the situation in India, as the Indian people have to live with it.

Note, for example, that for peasants in Third World countries, yield is not all that matters in the new HYVs. 'Miracle rice' was agreed almost universally to taste less good than traditional varieties, but then American agricultural researchers do not have to survive on rice as their staple food. The use value (as opposed to the market value) of rice and wheat to peasants often includes the use of stalks as thatch. The HYVs are dwarf varieties, useless for traditional roofing techniques. High-yield crops are not designed to work well – or at all – with others, which puts in jeopardy the widespread small landholders practice of multi-cropping: grains and legumes, corn and beans. This practice can increase both yield and profitability by more than a half, but HYVs put an end to it, and with it both a healthy crop growing regime and a large part of the protein in subsistence farmers' diet. Often the wheat would be sold and the legumes eaten. Now, with mono-culture and the market's discrimination against legumes, India's production of chickpeas has dropped from equal to around half that of wheat over twenty years, and the replacement of high-protein crops (legumes, around 20–30 per cent) by low-protein crops (grains, around 7–14 per cent) means that Indian people's intake of protein must have been significantly reduced in favour of carbohydrate.

Thus the transformation entailed by widespread use of HYVs carries implications which run deep into the structure of experience and everyday life and health in Third World countries. These are in no way simple and separable 'problems' of organisation or finance or research. They are to do with the particular coherence of the indigenous culture as a system of ideas, attitudes, practices and material things from houses to food. They are to do with feeling well and eating well and knowing what to do. If peasants do accept the gains from HYVs (which may be only quantitative) then it is despite the perceptible losses.

Gains in quantity are themselves highly conditioned, because the new varieties are not so much high yield as high response. There is much more to it than sowing seeds and waiting for them to come up.

Eastern India is the traditional rice-growing area. But it is prone to floods – and flooding drowns the short HYVs. The other side of the coin is irrigation: farmers growing the dwarf grains must have the right amount of water in the right place at the right time. Taken together it is a problem of water control, rather than irrigation in a simple sense, and water control costs money and is therefore in the hands of the big landowners.

The situation is highly ramified, because one cause of now-serious flooding year after year is forests in the Himalayas being cut for commercial and domestic use. In its turn this is related to peasant communities' needs for fuel, and because cowdung is often the main fuel, this in turn limits the amount of organic fertiliser reaching the soil. This means that levels of some trace elements such as zinc fall because they are not replaced by chemical fertilisers. And so on and on. It very soon becomes clear, to those who struggle with the practical obstacles to increase grain yield, that it is not a matter of 'certain problems' of whatever kind; it is a matter of the whole connected complex of *use* values.

Research might provide some degrees of freedom in such a situation, as in providing alternative second crops to replace flood-damaged main crops, or in developing methane gas (biogas) generators which release the calorific value from cowdung for heating and cooking while also producing an organic residue for fertilising the land. But most peasants are very poor, their holdings very small; and research too costs money. Not only must its place be carefully negotiated within the intimate economy of use values, at the level of village life, but also within the vast, impersonal, international economy of money and market values.

Another consideration leads to this same daunting conclusion. The farmers of Eastern India have to cope with flooding. But they also have smaller landholdings than in areas with more wealthy farmers such as Tanjore, where miracle rice produced its first crop of violent public disorder. Small farmers cannot afford expensive chemical fertilisers whereas rich farmers in the Punjab were able to give HYVs the conditions they need – high loading of fertiliser, careful water control, large areas devoted to single crops which are amenable to mechanised farming, pesticide to kill the pests which this encourages. The result is that the Punjab, not traditionally a rice-growing area, sold twice as much rice to the government in 1978 as did the whole of the rest of India.

Implementing the high-yield technique implied radical shifts in culture and the apparatus of production. The traditional rice-growing region became a lesser economic focus and its poor farmers became even more economically marginal. Village economy was disrupted as patterns of farming, marketing and wealth became more grossly deformed. Small farmers became even more insecure, not only from the susceptibility of HYVs to floods and suchlike but also from a generally greater variance from year to year around a higher average yield and from downward market pressure on the prices they can get for their surplus. Many small farmers were ruined, had to sell up, and became landless labourers. Rich landowners and many medium landowners made a fortune. Among rural families, twenty-two per cent have no land and forty-seven per cent have less than one acre. Six or seven per cent, among the richest landowners, live off the fat of their land without exploiting the commercial advantages of HYV technology. This leaves three or four per cent – rich *modern* farmers – to take all the financial benefit. The 'obsolete' social structures – patronage, private accumulation and financial speculation – thrive on the modern technology.

This is not a 'problem of finance' or of 'obsolete social structures'. It is a fact of life rooted in the uneven development of agriculture within a nation of profound and entrenched class, caste and geographical differentials, aggravated by imperialism. It is a problem of political and cultural struggle between groups with different needs and amounts of power. Techniques, organisation, finance and research are parts of the problem rather than external 'solutions' to it, because they compete for funds and because they systematically affect the relative power and effectiveness of different groups: landless peasants, farmers, other businessmen, research workers and corporations.

A ROCKEFELLER RIP-OFF?

It is an open question how far the original Rockefeller research programme and subsequent researchers' motivations have been determined by inter- and intra-national power plays. But there is no question that the developed situation in world agricultural research is manipulative and machiavellian in the extreme. The technicism of the Green Revolution (the idea that research on food grains might provide a technical fix for poverty) has been a Trojan horse for any number of commercial, national and elite interests in Third World countries. Consider, for example, the fact that spending on agricultural research by the Indian Council of Agricultural Research grew from Rs 64 million in 1965–66 to Rs 370 million 1976–77; yet it still remained less than the amounts spent on defence, atomic energy or space research. What does this say about the power of 'national' prestige versus the subsistence needs of the people who by convention make up the nation? What does it say about the power of different groups of research workers – elites who exploit internationally valued fields of Big Science and high technology such as space research, relative to researchers whose work might respond more directly to the needs of small farmers and low-caste peasants? These are deeply political matters intrinisic to the structure of science in the Indian state.

Consider too the needs which have been satisfied by an avenue of economic development which required huge quantities of chemical fertiliser. Thanks to the British Raj, India had a well-developed business class and a pool of technically trained white-collar workers, for whom a heavy chemicals industry provided means for advancement at the expense of the urban and rural poor. Thanks to the dependence on imported chemical plant and design knowhow thus created, the Indian economy and the livelihoods of millions of workers and peasants became a bargaining counter in world politics. The USA uses food aid as a lever on Indian internal and external policy, and as India's needs moved from grain to the means of producing grain (fertiliser and industrial plant), they became embroiled in a power play between two US agencies responsible for two types of aid. These too are deeply political matters, intrinsic to the structure of science and technology as an international system of wage work and commodities.

In the face of considerations like these the Green Revolution philosophy looks almost like wilful avoidance of the truth. Yet it has become a convention, to the extent that to get a research grant for plant breeding in the USA today it is almost obligatory to say something about possible outcomes for easing the world's food problem – even when it is not true – and the USA cannot get rid of all the food that it produces anyway, paying farmers to keep land fallow. More than a convention, many research workers in the area believe that they are personally doing something useful beyond the limits of their own careers and employers' marketing plans. Techniques can give some real if small autonomies to those who have little autonomy to begin with – though these are not necessarily the

42

techniques that are researched in labs in the First World. They might allow choices of crops and some flexibility of planting in exploiting otherwise unfarmable land, and so on. But technique itself does not have very much autonomy, hedged in by money, multinationals and entrenched class power.

As early as 1974 even the members of a United Nations research institute could draw this conclusion. In projecting First World thinking about technology and its social structure into the Third World the Green Revolution showed how green the technicians were. Forces of inequality and forces of nature beyond the comprehension of jigsaw-puzzlers crushed simple notions of 'feeding the world's poor'. The UN institute put this in a characteristically diplomatic report. The criticisms were ignored because, even in UN bureaucratese, they were devastating and unanswerable.

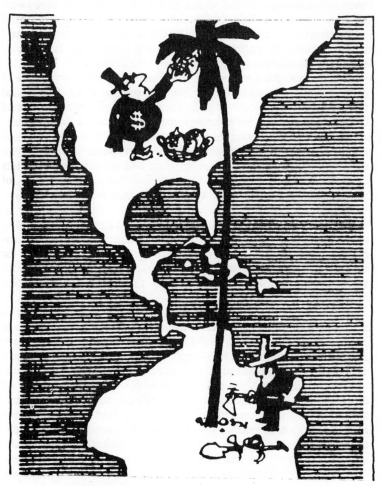

Since then the situation has become even more complex and explosive. In its present phase the Green Revolution is becoming part of the business of large multinationals who have moved in on plant breeding. This is hardly surprising, since the majority of research workers in the capitalist world work for these firms. But it does put new twists in the tale and it makes speculations about machiavellian motives more convincing.

Multinational companies made and sold the fertiliser for the Green Revolution (Rockefeller owned Esso which pumped the oil which made the fertiliser which raised the yield which ruined the farm that Ajit built). They made and sold the manufacturing plant when India wanted to produce fertilisers, and they kept the design skills and the spare parts. Multinational companies made and sold the pesticides which large scale cultivation of HYVs depends on. But all of this business was within the established boundaries of industrial production in the First World. Now the agribusiness interests are moving on from their bases in research and marketing, to diversify into the seeds business, or rather, seeds are being made into business. New legislation now being passed in First World countries is giving patent-type protection to new varieties of plants developed in the research and development laboratories of the multinationals. The companies are accumulating seeds as capital, by creating privately owned 'gene banks' of seeds out of which to produce future new varieties for sale. One company, United Brands, is reputed to hold two thirds of the world's stores of germplasm in banana species. United Brands used to be called United Fruit, and under that name became notoriously linked with CIA plots to depose reform-minded presidents in Latin America.

Some public gene banks have had difficulty in persuading private companies to tell them about the kinds of genetic material they hold, or how much. Capital is capital even when it is seeds, and business is business. 'Bugs' technology (the splicing of genes to produce new genotypes to specification) means that over the next decade a widening field of laboratory-made plants will be exploited commercially, and with this comes the possibility of many farmers' complete dependence on non-natural varieties owned by multinational companies. Farmers using HYVs must buy seed anew each year because, as hybrids, these varieties are sterile. But in addition, rice growers in the Philippines have already found that after suffering disastrously from the failure of a series of 'miracle' varieties they are now locked-in to the seeds market. The original, robust, pest-resistant indigenous variety is extinct. In future others may be 'lucky' – they may be able to buy back their original strains, at the price of continued dependence, from the multinationals' gene banks. Whatever the technical future holds, there is now pressure on the UN to oppose present legal moves by declaring plants as 'resources
44

of common heritage to all peoples' and thus non-ownable. There is, of course, pressure from the seed companies to retain the *de facto* right of ownership.

Even the material that plant scientists work on has now become overtly political, by becoming a commodity. What price the narrow technicism of the Green Revolution as a vision of progress? Did it escape your notice that many of the Vavilov centres of genetic diversity are in countries where imperialist and anti-imperialist struggles have made them into centres of war?

SCIENTISTS AND THE PEOPLE

Since this chapter is about public policy, or more accurately, the relationship between corporations, states and people, there is a sense in which it and the preceding one are about science policy. But the sense is far removed from what that term usually implies. The Office of Science and Technology Policy is part of the White House apparatus which feeds lines to the US President. The Office of Technology Assessment is part of the US Congress establishment. In Britain, units for science policy and technology policy exist in various academic contexts and in 1980 a Technical Change Centre was set up, financed by government research councils and capitalist conscience money (the Leverhulme Trust – Unilever is a massive corporation based in Third World food resources) to develop 'a major programme of research on the choice, management and acceptability of technical change relevant to the advancement of the national economy'.

Science policy, as an activity of researchers in such institutions, is a kind of comparative economics which produces statistics to show how much better we need to do in quantitative comparison with our economic competitors. Some researchers and some units try to do more, but introducing social considerations into this official-oriented discourse is an uphill struggle. Money talks. This chapter and the preceding one have been about politics rather than economics, and ends rather than means or measurements. As a basis for further discussion about science politics – rather than science policy – there are some broad topics which may be drawn out from the two chapters.

Neutrality of Scientists

Science is not a thing, but scientists working on things, so that 'the problem of science and society' necessarily has to be reformulated as problems of scientists' relationships with other workers. At least three levels can be seen where problems crop out in the context of cultural and political change. First, the 'jigsaw puzzle' level. Scientists get hooked on problem-solving, that is, on using what they know to put together a picture which they already have the key to. The internal logic of their models and formulations and research programmes keeps their heads down. This is certainly a factor in the Green Revolution's conservatism and maybe in the Soviet geneticists' conservative response to an economic and political crisis. The fixation with known methods and known technical problems in normal science is a cultural underpinning of the technical fix as a political technique.

Second there is funding. Scientific and technical practices require resources, often expensive, and this constraint tends to put a premium on clearly defined and orthodox specialisms, on 'normality' and quantity (more) rather than quality (different and better). And third, there is patronage. In science as a career within a corporate structure this constraint is 'normal'. But in Lysenkoism it was dramatic. Power games within the Soviet scientific and Party establishment, ending in ruined careers and even some deaths, were a grotesque parody of career competition within 'normal' science. Taken together these three areas of consideration make the conventional idea of scientists as 'neutral' very problematic. Researchers are in fact committed and have their own interests, and while these may not always be political in any obvious sense they clearly have a bearing on the roles that science will play 'naturally' in political affairs.

Technical Fixes
Both Lysenkoism and the Green Revolution illustrate some limits of technicist voluntarism. There are social limits, so that projected technical 'solutions' to problems run aground on culture as a material organisation of people, things, places and money. Lysenko's proletarian science, even if it had worked, would not have raised the grain output of the USSR because the cultural unity of agriculture and industry, country and city, had been shattered by the vulgar materialism of a philosophy of which Lysenkoism was an integral part. As a particular type of science, Lysenkoist genetics depended on the suppression of culture both in theory and in practice. Only then could technicians feel that they held the key to social progress. Rockefeller's unproletarian science operated with a less arrogant variant of the same notion, that cultural structures are less 'real' than technical structures and that therefore getting the techniques right – in an abstract, laboratory, First World context – is what matters. The rest is just 'problems' that need peripheral tinkering, more money, and teaching peasants to adapt.

There are also natural limits which mean that when the laboratory tested 'solutions' are run out in the field they may not in fact work, even at a technical level. Arrogance and urgency combined to prevent Lysenko's solutions being even laboratory tested, and nature's stubborn intractability soon showed the weakness of the theory's narrow intellectual base. That base might have been wider of course, if Lysenkoism had not also required a refusal of the possibility that orthodox scientific principles might be in some sense 'necessary'. Green Revolution science, on the other hand, was tested and shown to work in trials. But in the large-scale system of other plants, pests, predators, climate and land use, the narrowly engineered 'solutions' of technicians fell through. It might be argued that agricultural science is in principle perfectible in respect of its ecological sensitivity. But it is hard to see how that could be, when narrowness is the very mode of investigation which gives modern science its power and productivity. Another science might work, in a total sense, but not an extension of *this* one. In order to see them as solvable through mere technique, an approach to problems in the real world must separate them off from the lives, the cultures, in which they become problems rather than just phenomena.

The differences between Lysenkoism and the Green Revolution as technical fixes are worth noting. Rockefeller science was a plain technical fix which turned out to be politics when it was let out of the laboratory. It was politics – the politics of seeming to be above politics – which enabled it to masquerade as technically self-sufficient in the first place. Lysenkoism on the other hand was an overtly political fix working through a technical fix – which happened to be no fix at all but that's incidental, since Lysenko might just as well have been a brilliant scientist. It still turned out to be politics of the most unpleasant kind for the scientists involved. The problem is not that politics and science were mixed, since sciences are always part of the power structure of societies, whether marginally or centrally. The problem in both cases was that the politics was bad politics. The courses of action chosen in Lysenkoism and the Green Revolution cut across real existing social and economic structures, pretending they were not real. Whether politics is for or against the *status quo*, it is essential for success to take the whole of reality seriously, considering the relative entrenchment of all major conflicting and contradictory social practices.

Playing the politics of the technical fix can drive the ideology of 'neutrality' to quite fantastical lengths. In the USSR it was accepted, at one level, that scientists were not neutral in class terms, and that the class struggle is active within science just as elsewhere in culture. Yet this explicit political recognition was implicitly negated at a philosophical level by the redefinition of those who opposed the revolution as 'mistaken' and those who supported it as 'correct', that is, scientific. Marxism, as Diamat (dialetical materialism), was defined as a science. Hence the scientists who opposed were 'really' unscientific and so science (Diamat) became invoked as the 'neutral' arbiter in political practice. The double blind involved an overtly political fix which then turns out to have been 'scientifically' correct.

Proletarian science on the grain front placed itself in the context of revolution. Unproletarian science has to be seen – but will not see itself – in its place as part of the long revolution in industry, culture and democracy which 'modern' life has at its centre. From the institutional violence and from the sophistication and sophistry of technicians' approaches which can be seen in these two cases, it must be clear that the relations between science and revolution are more tangled then we understand. But understand them we must, because they otherwise will not change. Or at least, there is no guarantee that changes will be for the better.

Development and Waste

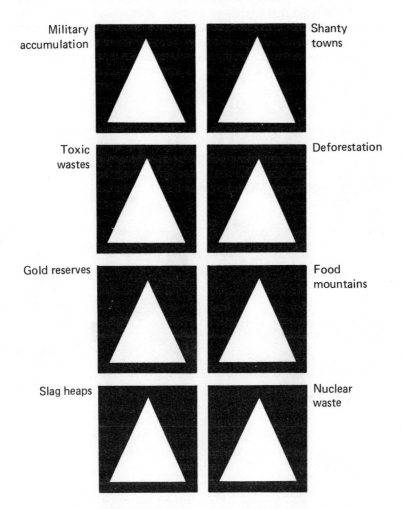

Military accumulation

Shanty towns

Toxic wastes

Deforestation

Gold reserves

Food mountains

Slag heaps

Nuclear waste

'Development' is a loose and often misleading term. Israel and Egypt are 'developed' within the context of the Middle East. The Seychelles are 'developing', but not in the same way as the Falklands/Malvinas will develop. Ireland is underdeveloped within Europe, as is the Mezzogiorno in Italy. Britain is underdeveloping as an advanced industrial nation.

Seen specifically as capitalist development, such isolated cases appear as elements in a process. This economic, technological and cultural process operates on a world scale and has among its effects many which are usually seen separately and called 'waste'. In this section we picture some of these as residues of a single, complex developmental process.

The residues take many forms. We show eight of them as 'mountains' which are currently under development, in a world landscape of needs.

Beautiful Eire

Top: mountains of arms, a first stage of development within imperialist rivalries, and a means of capitalist financial accumulation.

Bottom: exploiting the underdevelopment of Eire as 'charm', investors also supply the land with waste dumps and poisoned waters.

Top: gold reserves, mountains of pure market value, are an oppressive form of waste to those whose dangerous labour serves to replenish them.

Bottom: the 'Cornish Alps' are spoil heaps from clay workings. Steel and coal workers often live in a landscape of waste generations old.

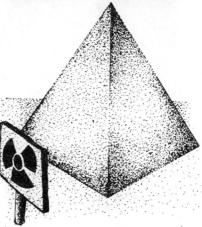

Top: homes built from the
wastes of cities accumulate in
suburban wastelands, as people
are laid waste by landlessness.

Bottom: millions of tons of
Himalayan topsoil, exposed to
rains, now accumulate as a silt
mountain in the Bay of Bengal.

Top: In Europe, butter mountains
and wine lakes. In the US, hogs
slaughtered and land fallow under
federal grants. In India, grain
'surpluses' and starving people.

Bottom: Egypt has offered to
take nuclear waste in exchange
for militarily exploitable energy
technology. Will they build
pyramids over it?

4   Scientific Workers

STRUGGLING ALONE with some rarefied aspect of nature: this is our most common image of the scientist. The relationship between scientists, and between scientists and the rest of society, as well as between scientists and the objects of their investigative labours, has in fact changed materially over the last three centuries. This is underlined when we think of perhaps the second most common view of science which comes into our homes – the wide open spaces of Mission Control at Houston, filled with screen-watching, knob-turning people. Are those people scientists, and where does science fit into the Barnum and Bailey world of space shots? The white-coat pioneer image is a selective rendering of some outcomes and stages of the process through which sciences have developed. The technicians at Houston Space Centre, on the other hand, are the tip of an iceberg of present day scientific work. By looking here and in the next two chapters at science as a kind of work within the deepening division of labour in society, we may perhaps be able to see from where the limited and confusing validity of these images of science derives.

PRIVATE MEANS AND SCIENTIFIC ENDS
Up to about a hundred and fifty years ago scientists were not

HEADPIECE The laboratory of Justus von Liebig at Giessen, Germany, in 1842. Gentlemen wore hats – and only labourers had to take off their coats. This was the first teaching laboratory, and Liebig's pupils became the professors of the internationally pre-eminent, industry-serving German chemical establishment. (Lithograph: Liebig museum, Giessen) workers. This means, for one thing, that they did not earn their living through science. When Isaac Newton was given a university chair at Cambridge in 1669, at the age of twenty-six, professors were not expected to do any research, which is why a young man with no published work to his name could even be considered for such a post. Later in life he got £400 a year as Master of the Mint, but again, researching was not part of what he was paid for. Newton's older contemporary, the Hon. Robert Boyle, was the son of a wealthy Elizabethan land-grabber, first Earl of Cork and owner of considerable interests in spinning, weaving and ironworking in the contested territory of Ireland. Boyle wrote that, 'My condition does, God be praised, enable me to make experiments by others' hands.' Patronage and private means were the life blood of science, though it caused the death of René Descartes, the French mathematician. His royal patron, the Queen of Sweden, could only see him at six in the morning and as a consequence Descartes caught pneumonia and died of it.

Those few who were able to earn a living through scientific work, because they had no means of support other than paid work, depended upon the good graces of wealthy patrons. Robert Hooke, for example, became attached to Boyle as his 'other hands' after meeting him at Oxford where he was a servitor. Boyle, whose family had itself only recently ascended the social scale, rejected traditional aristocratic values and supported the Commonwealth following the Civil War because, 'In republics the way to honour and preferment lies more open to desert, which is a quickening spur and a great incitement to noble spirits.' They certainly were not doing science for the money.

Although Newton rose to be able to dispense patronage himself he depended at first on the favours of a political patron, the Chancellor of the Exchequer. As with the arts in general, work in science throughout the eighteenth century remained within the sphere of private patronage. Wealthy men carried out experimentation as enthusiasts, and those like instrument makers, mechanics, physicians or clerks, who had to earn their living by working at other occupations, pursued their experiments and discourses in their spare time. This meant that scientists were mostly gentlemen, of modest means.

By the beginning of the nineteenth century another possibility developed: the pursuit of a career in a 'position' financed by public subscription. Humphrey Davy (described by J. D. Bernal as a 'great scientist but even greater snob and showman') acquired such a position as first director of the Royal Institution in London, founded in 1799 through the efforts of Benjamin Thompson, Count Rumford of the Holy Roman Empire. Thompson was an American Tory whose interest in theories of heat was stimulated (while managing the affairs of the kingdom of Bavaria) by the practicalities of boring cannons. Lectures at the Royal Institution were too popular for the tastes of the gentry whose purses supported them, and mechanics became barred from entry – the doorway was actually bricked up. This reflects the general class affiliation of scientific work in the nineteenth century. Michael Faraday, assistant to Davy and later famous for his electrical researches, was an exception, having first been a bookbinder's apprentice. The employment of Davy and Faraday incidentally created a centre of excellence in science, while reproducing the dependence of research on patronage.

Jobs for thinkers and experimenters slowly multiplied through the 1800s, although Charles Babbage and others felt impelled in mid-century to broadcast doom-laden messages about the 'Decline of Science' in England, compared to other nations. In 1851 Lyon Playfair made the shocking discovery that in Britain there were less than 1200 salaried posts in science, philosophy and literature. The Arts and Sciences were at that time regarded as similar for administrative purposes. Help came in the form of the Great Exhibition of 1851 which bequeathed three new science education institutions in Kensington which later formed the basis of Imperial College. Scientists gradually got jobs in schools and universities where practical involvement in research, especially in newly created colleges, acquired increasing importance. They also found employment in government laboratories and eventually in industrial research laboratories. By the end of the century, most scientists drew salaries from universities or the government rather than having to depend on direct, personally negotiated private patronage. They had become 'gentlemen' in the much modified sense of lower-middle-class people with a position and prospects, different from the gentlemen of modest but independent means of the early century, and different again from the aristocrats, gentlemen and hangers-on of cultivated society in the eighteenth century. Scientists were now paid to carry out research. They were workers, though 'professionals' is the name that middle-class knowledge workers choose to dignify their labour. The work progressed within a career hierarchy ordered under professors and senior scientists. Pursuit of knowledge was systematically located within a growing body of factual knowledge and theory, and this progress was itself due to the parcelling-out of research tasks within a growing division of experimenters' labour. When science had depended on private means its purposes had been, in a significant sense, private purposes. The same could no longer be said of the corporate organisation and goals of science at the beginning of the twentieth century.

THE COMMUNITY OF SCIENCE
The now common idea that there is an international 'community of science' arose in connection with the very visible productivity of natural scientific research in the late nineteenth century, though in

itself the idea was a reworking of an ideal that was current three hundred years before at the time of Francis Bacon. Work within sciences is in fact hierarchical, fragmented and generally routinised, so there can be little meaning in the reference to 'community'. Given that the productivity and the international organisation depended on the fact that science had become a form of *work* within capitalist societies, there is no reason to expect that 'community' is a useful term for understanding it.

The term 'scientist', replacing 'virtuoso' or natural philosopher, was coined in 1840 by William Whewell. But by the time that it had been in currency for a generation it was already problematic in the way that it implied (as Whewell had intended) a cultivator of science *in general*, for science became more and more *particular*. There are today, and have been for a century and more, many kinds of scientific work. There are sciences – physics, chemistry, biology – and there are inter- and sub-sciences like biophysics and biochemistry, metallurgy and genetics; in addition there are recent empires like molecular biology that stand aloof from traditional ones. Specialists in particular fields generally do not communicate with one another and it may in fact be strictly impossible for them to do so by virtue of their specialism. Each field has a sub-cultural identity. Physicists are eccentrics, much as the landed aristocracy were after the industrial revolution. Chemists are 'bangs and stinks', engineers are beer-swilling *men*. The point is not whether these images are accurate, but that they are totems of tribalism in technical life, signalling real obstacles to communication, cooperation and a wider relationship within a popular culture.

It is doubtful even whether members of a specialism talk effectively to each other when, for example, not only books, not only journal articles (a briefer form of publication), and not only 'letters' journals (invented as a still briefer form, to improve on the circulation of articles), but also a pre-prints network exists which, in a single subfield of physics, may shuttle hundreds of items a week around the 'community'. Five, maybe ten, thousand papers a year. Who reads them? A scientific research worker only reads within a very narrow subfield, and then selectively. The average scientific paper is read by less than two people. In chemistry there are more than two hundred and fifty journals, and for the top twenty-five (those most cited in 1978 and '79) the average rate of citation is 2.2 times per paper. *Physical Review* found it necessary, over the space of a dozen years in the 1960s and 70s, to split into six parts, each larger than the original journal. Is it surprising that a research worker or library may stop subscribing when the shelf space runs out or when spending cuts come home to roost?

Science is also split by spheres of employment: universities, polytechnics, private or state-owned firms or state-supported research institutes and agencies. Some scientists are in private business as consultants. This is happening especially in the biotechnology area where consortia of senior academics are going out on the market looking to sell their research groups' know-how. Research is, of course, split by geographical distance and international cultural and political divisions, so that the material reality of community, meeting face-to-face, is determined by access to funds for air fares and conference fees, which is in turn a matter of seniority and patronage. Research, as work in a hierarchical society, is itself split hierarchically. Many scientists do not do research at all, either because of their lowly positions (students, lab assistants, teachers – as passers-on of others' accumulated wisdoms) or because of their committee-sitting time-economy as the top management of the scientific establishment (professors, research managers).

As in all corporate spheres, a scientific worker generally has to get promoted out of the job in order to 'get on' – into management and often out of competence. There are many gradations within 'middle management', marked by labels such as post-doctoral fellow, section leader, assistant or associate professor (the US equivalent of 'lecturer' and 'senior lecturer' in Britain), senior development chemist, research assistant. Very few of these workers have any autonomy to pursue individual purposes in their work as

---

GETTING ON

THE CORPORATE STRUCTURE of science means that either you get locked into a job at one level or you get promoted out of the job altogether. This writer refers to computer programming and teaching, but the picture is much the same in any 'professional' field:

> From team leader, the worker may go on to become chief programmer, project leader, or head of a division ... Whatever the actual steps, there is likely to be an increasing element of management and less and less technical work. (I believe that teachers find the same difficulty, in that to make career advances it is necessary to leave behind what they want to do, i.e., teach, and go into administration.) Many programmers reject this path because they want to remain in programming. But after some 3 or 4 years they will find themselves being questioned closely at interviews about how much actual 'responsibility' they have held. It is made clear that 'responsibility'

does not refer to programming work but to the number of other people they can claim to have controlled. Eventually they will find difficulty in changing jobs, because by now they will, through annual increases, have reached a salary level considerably beyond that of a programmer with one or two years of experience. A prospective employer is likely to see two such applicants – one with 7 or 8 years' experience and one with 1 or 2 years' experience – simply as two programmers, the only difference being that the former is very much more expensive than the latter....

The end result, then, is that anyone who has not climbed the career ladder is considered to have missed the boat. Since it is clear that only a minority can fit on the ladder, the result is a denigration of those who remain behind – a denigration which would not exist, or not in the same form, if the ladder did not exist.

(Ellen Crowe, 'The Loneliness of the Long Distance Programmer', *Radical Science Journal* 11, 1981)

'scientists', though they do pursue individual careers within the hierarchy. The conditions are those of work, that is, wage work, and the community of science is the most highly refined sector of an international division of social labour.

The Socialisation of Science
We can talk about this by saying that science is now highly social-ised, in a special sense. This means that it has a refined, divided and extensive internal social structure. It also has an intrinsic and multiple connectedness with almost all other spheres of social activ-ity: government, investment, education, consumption, employ-ment. The story of the socialisation of scientific knowledge seems paradoxical because it is in fact a story of fragmentation, but the paradox has to be understood and explained through seeing the history of science as a history of work within an industrialising culture. Science became work, in the sense of a paid occupation, during the same period as work in this same sense – wage work – was becoming entrenched as the dominant condition of making a living. The last third of the eighteenth and the first decades of the nineteenth centuries was the era of the industrial revolution in Britain. It was the period during which scientific positions began to be established within the division of labour. The late nineteenth and early twentieth century was the era of imperialism, the period during which mass employment in research became a significant condition of international competitiveness in manufacturing indus-try, as part of the growth of firms and of state administrations. Management and engineering as professions, the social sciences, all have their origins here.

Modern science as thus constituted is a very powerful mode indeed. It has to be stressed that we know no other science. The values of 'science' as we inherit them today are values of work of this particular kind.

This means that if there is to be any revaluing of science and its ends this must proceed through a revaluing of its forms as well as contents and priorities: hierarchical, fragmented work, a body of practices and interests which as a whole are distanced from personal negotiation even for a member of the 'community'. All the status that we now attach to research as a national and human priority is associated with the particular productivity of this mode of organising intellectual work. Could there be other modes? Why not?

SCIENCE AS DIVIDED LABOUR
The mode is not accidental, though it has been a long time coming. In his *Great Instauration* of 1626, Francis Bacon anticipated that 'Many shall runne to and fro, and take paines in finding it out, and Knowledge shall be increased.' As a project this was taken up and embodied, most notably, in the aims of the Royal Society of London

for Improving of Natural Knowledge. As an organisation the Royal Society did not succeed in turning experimentation and research into useful products and processes. That general and systematic incorporation of science in technology awaited the fuller development of industrial production itself in the nineteenth century. The continuity of Bacon's ideals through the Royal Society lay more immediately in the images that they offered of how scientific work and society generally might be ordered. As a group the Society was forming over the period 1660 to 1663, when its second royal charter was issued, and among the founders with Robert Boyle, the Irish landlord, was William Petty, chief physician of the Cromwellian army which had bloodily suppressed a rising of Catholics in Ireland. Rewarded with Irish land, Petty found this an ideal opportunity to put Baconian principles into practice and he proposed, and was commissioned to carry out, an ambitious survey of the whole of the island.

A VOYAGE TO LAPUTA

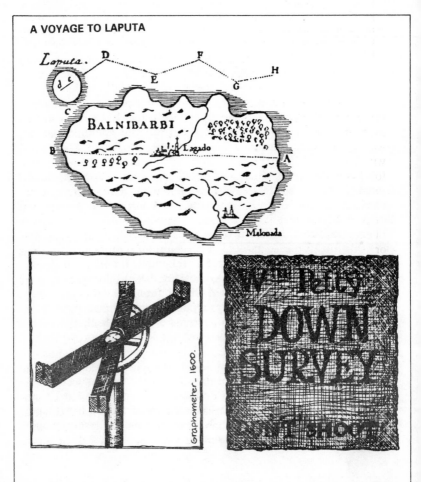

IN GULLIVER'S TRAVELS: 'A Voyage to Laputa', Jonathan Swift satirised the Royal Society of London and the Baconian project of useful knowledge. Laputa, a floating island, is a scientocracy which governs the land-bound island of Balnibarbi. Gulliver descends from Laputa to visit the Grand Academy of Lagado and meets a man:

> of a meagre aspect, with sooty hands and face, his hair and beard long, ragged and singed in several places. His clothes, shirt and skin were all of the same colour. He had been eight years upon a project for extracting sunbeams out of cucumbers, which were to be put into vials hermetically sealed, and let out to warm the air in raw, inclement summers. He told me he did not doubt in eight years more he should be able to supply the Governor's gardens with sunshine at a reasonable rate. . . .

Now where have we heard that one before? Something about extracting sunbeams from . . . atoms, wasn't it? – to supply the people with energy at a reasonable rate, once the Governor's needs were satisfied.

A tailor came to take Gulliver's measure for a suit of clothes. But . . .

> This operator did his office after a different manner from those of his trade in Europe. He first took my altitude with a quadrant, and then with a rule and compasses described the dimensions and outlines of my whole body, all which he entered upon paper, and in six days brought my clothes very ill made and quite out of shape, by happening to mistake a figure in the calculation. But my comfort was that I observed such accidents very frequent and little regarded.

The Down Survey, as it was known, was executed by foot soldiers of the New Model Army under Petty's direction. His aim, expressed in terms directly echoing Bacon's description of scientific method, was to supplant 'persons of gentle and liberal education equipped with complex rules and instruments', as would normally be required for a sophisticated technical task of this kind. Petty redesigned instruments so that they could be used with minimal training for particular defined purposes and capitalised on the rigid hierarchical division of labour in the New Model Army to implement the survey under the supervision of a limited number of skilled surveyors. The organisation of the Army was itself no accident for Bacon's writings had acquired the status of moral tracts within the ranks of Puritan opponents to traditional rule in the Civil War. Foot soldiers were paid piece-rate, by the linear mile, to encourage them to face rough terrain and hostile populations, while supervisors were paid an annual salary to discourage short-cuts. Petty himself was remunerated according to the *area* surveyed.

Having visited Holland as a young man, Petty was inclined to attribute Dutch trading superiority to the highly developed division of labour in Dutch industry. In his *Political Arithmetic* of 1682 he proposed, a century before the better known description of pin-making which opens Adam Smith's *Wealth of Nations*, that a rigorously simplified and divided labour process might produce a watch 'better and cheaper than if the whole work be put upon one man'.

The Baconian Ideal

The connection with Baconian images of sciences is direct. Bacon stressed, first, that science should be the coordinated activity of numbers of individuals, in contrast with the individualistic and isolated practices of alchemists and mathematicians working in the Renaissance tradition. 'All works are overcomen . . . by conjunction of labours,' he argued, so that whether or not a modern scientist possessed genius to match the ancients, he might as a duly disciplined member of a collective labour overtop all of their knowledge. Noting it as a fundamental condition of productivity in capitalist manufacture, Karl Marx was later to name this the principle of Simple Cooperation. Secondly, however, 'soundness of direction' mattered to Bacon; that is, the work of all had to be overseen by a few. In his fantasy of the House of Solomon three scientists were empowered to 'direct new experiments, of a higher light more penetrating into nature than the former'. This meant, thirdly, that each individual worker had an individualised task, defined through 'a form of induction which shall analyse experience and take it to pieces', each piece being the basis of an isolated individual's labour. Hierarchy and fragmentation together meant that the conception of tasks was functionally separate and socially separated from the carrying out of tasks, in just the way that Petty's foot soldiers were mere measuring instruments with legs. The effect of such task-discipline, finally, was social order. Bacon claimed in his *Novum Organum* that the method of divided collective labour in science 'performs everything by surest rules and demonstrations', stressing 'how strict and disciplined a thing is research into truth and nature'.

Such principles were not, of course, Bacon's alone. Petty had seen them in practice in Holland, where they had in turn been gleaned from the fearsome discipline of Calvinist Geneva. Bacon's method, Petty's survey, Calvin's Geneva or Cromwell's army – what we see working here is a radical principle of social organisation which, through the agency of just those social classes active in the Royal Society, became in the eighteenth and nineteenth centuries a general mode of social labour and, eventually, of scientific labour too. The mode is clearly far from accidental. But it was a long time coming, and its coming is part of a very long revolution, which has as constituent phases and aspects the Puritan rebellion of the 1640s, the anti-radical Glorious Revolution of the 1680s, the scientific revolutionism of the Royal Society and the industrial revolution of the late 1700s and early 1800s. It keeps on coming.

What this means for life in the modern 'community' of science was sensed uneasily by C. H. Waddington, a biologist and a Fellow of the twentieth-century Royal Society, writing in *The Scientific Attitude* (1941): 'Science in general has tended to become an enormous collection of details', he sadly explained. 'Any individual may feel a certain justifiable pride if he knows that he has added one brick to the structure. But in recent years, more I think than in the past, it has been assumed that the discovery of one or two nuggets of knowledge is all that a scientist need attempt.' The writer was mistaken in looking back to a golden age. Before science was incorporated, and before its labour was divided and partitioned between detail-labourers, it was not science. It was what today we call, condescendingly, 'prescientific'. The narrow division of labour, and the narrowness of scientific aims, is strictly intended as the norm. In contemporary research in molecular biology, for example, the animal chosen for research (say a nematode worm) is carved up into researchable bits, the gut to one worker in a lab, the head to another, the tail to a third and the muscles to a fourth. There isn't that much of a worm to go round, and I have friend who came into the system a bit late. He got the eggs to make his name with. This is called team research, and in a normal (hierarchical) group only team leaders have the basis for studying how the bits go together and planning future exploration. This is the 'community' of science.

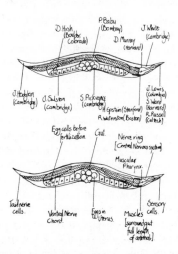

THE SOCIALISATION OF SCIENCE, AND NATURE AS A MAP OF SOCIETY
Above: the carve-up of a nematode worm among members of an international research 'community'. Below: physiological divisions of the animal as an object of labour in molecular biology.
Nematode worms are mainly parasitic, like the roundworms of pig and human guts, but this was not a motive in choosing it as an animal to work on. The aim was to prove that the molecular biology approach works on this particular kind of stuff, and thus to reproduce a particular mode of working within the career structure of academic science. The eggs researcher, Simon Pickvance, explained: 'The process of development is a continuum and I was expected to look [only] at particular parts of it. There are processes with long names, like blastulation, gastrulation and so on, which normally form chapter headings in books because they normally involve different sets of workers' (' "Life" in a Biology Lab', *Radical Science Journal* 4 (1976), 11–28). (Drawing: David Young, *RSJ* 4)

SKILLED PROFESSIONALS

'Professional' is a dangerous word – it means so many slippery, important things. One thing it names is a skilled person who carries out a kind of activity for money, where once amateur practice – enthusiasts' practice – was the norm. In this sense science today is professional, in that scientific workers earn their livings by it. Yet that same word carries the implication that earlier work was 'merely' amateur. Newton, Boyle, Descartes – all amateurs! Who do professionals think they are? Another thing that 'professional' signifies is a person whom we are supposed to be able to trust with our interests because they know so much more about something than we do. To get an idea of just how much scientific workers do have to know, let's look at a relatively new 'profession', Operational Research. It's a useful case because its history is short but serves to illustrate a general proposition about skill and about knowledge in scientific work.

Operational Research

Operational Research – almost always known as 'OR' – was a discovery of World War II; the term was first used in 1939. Scientists became involved at a technical level in projects such as the development of radar and, later, the Manhattan Project which led to the Hiroshima and Nagasaki bombs. Some scientists, mainly physicists together with mathematicians and biologists – many of them either eminent or promising research workers in their fields – went further and involved themselves in the investigation of operational problems such as how best to use radar, how best to deploy limited resources such as airplanes, and how best to evaluate the effectiveness of different bombing strategies. This was OR, and by the end of the war it was widely felt that as a significant innovation it was worth expanding into peacetime non-military contexts. The eminent and promising research workers, however, went back to their labs, leaving would-be leaders of a new profession with a problem of recruitment and definition. Without bona fide scientists, certified as such by the approval of their peers in natural-scientific research, it was hard to establish a new mode of scientific research on the territory beyond the laboratory walls.

The practical challenge gave rise to much 'philosophising' about Scientific Method in OR and many past practices were ransacked for models of style and content, starting with the wartime work and other work in various areas of applied statistics (such as studies of road traffic flow and textile quality control) then trailing back over the centuries from Thomas Alva Edison through Charles Babbage and Isaac Newton, right back to Archimedes. By the mid-fifties members of the would-be profession had begun to hit a successful note. Addressing the Operational Research Society in London after a decade of post-war attempts at civilian OR a speaker opened his talk with:

> Operational Research is still young and self-conscious. It is, therefore, very much concerned with what it is . . . whether it is a science at all. This latter question has probably been settled at last, to the satisfaction of most of its adherents. It is agreed that it is a science. . .

The stuff about 'science' over with, he continued: 'I may assume, anyway, that my audience has already heard enough of it,' and he spent the rest of his talk weighing up the workability of a whole range of *techniques*, proposed and actual.

He looked at linear programming (a mathematical method for finding the least-cost combination within a set of constraints) which was judged useful, and game theory (a mathematical framework for playing-through competitive situations) which was judged OK for wartime but not powerful enough for the conditions of business OR. The theory of queues (which deals with the availability of services at points of demand, in factories, shops, railway stations or tollbooths) looked pretty effective, and a whole bagful of statistical techniques for processing number into information were emphasised as the OR worker's stock in trade. Standard techniques are what the 'profession' came to be built on, in simple imitation of the 'body of knowledge' which, as everyone knows, all real sciences possess.

Teachable Knowledge

In Britain it was almost another decade before the technical know-how of OR gained the status of a university chair (at Lancaster University in 1963), and in fact it was in the USA that the body of techniques was most rapidly developed. As early as 1950 a New York correspondence college had priced basic OR skills at $32.00. This rapid transition from scientific skill into marketable commodity was well in line with the much closer relationship which existed and still exists in the USA between military, academic and commercial practices. Robert McNamara, one-time US Secretary of Defense, illustrates this well. From operational research during World War II he moved as president to the Ford Motor Company, which had been a big supplier of military equipment. Then, via the Defense Department, he ended up as head of the World Bank. Along the way he orchestrated the financial control of 'miracle grains' research through governments, UN agencies and private foundations. He was an old Ford Foundation hand himself; most of the international crop research station directors have come up through the foundations.

Mathematical techniques in American business were well developed before the war and OR – as a wartime innovation – was appropriated within this established culture. This was especially true in queueing theory, game theory and linear programming techniques. The sophistication of US mathematical work in OR burst

63

upon the British OR community, whose technical vocabulary was mainly in statistics, as late as 1957 when the first international conference was held. Before long the journals began to look suitably mathematical and therefore scientific, and that pattern has become universal. To enter the profession nowadays a person becomes a student, not of 'science' in any general sense, but of OR. There are by now plenty of professors to teach them. The student becomes versed in a whole panoply of techniques and standard 'models' of

---

A 'MODEL' OF STOCK-HOLDING COSTS

A very well established form of OR model deals with the managerial problem of choosing the size of orders for goods which are to be kept in stock. The larger each order the more capital is tied up in stock, and so the stock-holding cost rises as the order size rises (line A on the graph). On the other hand, the larger each order the less often ordering has to be done, and so the cost of ordering stock over a whole year falls as the order size increases (line B on the graph). Adding together the holding and ordering costs gives a total which falls and then rises again as order size increases (line C on the graph). The least-cost order quantity is the one which gives the lowest point in line C – in this case, an order size of 3,000 items.

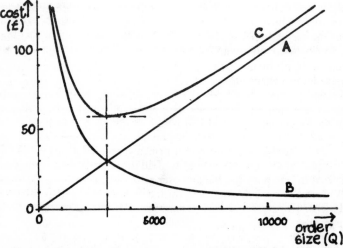

To make a mathematical model of this situation OR workers assume that goods are used from stock at a steady known rate. On this basis the least-cost order quantity, Q, can be calculated from a formula:

$$Q = \frac{2 U O}{C}$$

where U is the total usage of items in a year, O is the ordering cost per order and C is the cost of holding an item, averaged over the whole year. This formula and the mathematical statements used in deriving it make up the formal, algebraic part of the model. The definitions of what the symbols mean, and the assumptions about usage rate and so on, enable real-world knowledge to be coded into model-world 'language' – that is, algebra. They also enable model-world 'results' to be decoded into predictions of real-world behaviour: they are therefore an integral part of the model. Using the model, no graphs need be drawn. By slotting in a set of numerical values

64

for U, O and C, a value for Q can be calculated. In the above case Q turns out to be 3,000 – the figure given by the low-point of the graph.

The model can be used to assess the effects of changes. If U were doubled (twice as much usage in a year) Q would increase by about one and a half times, rather than doubling as you might expect. If the guesses of future values for U, O and C were inaccurate this would affect the accuracy of the prediction. However, reworking the mathematics indicates that a Q that is too big is less costly than a Q that is too small – so that when in doubt about the accuracy of figures it would be more economical to err on the big side in choosing an order quantity. Any mathematical model can be used in this way, not just to churn out a single 'best' answer wholly dependent on assumptions, but also to indicate how sensitive that answer is to the reductive assumptions that must always be made in turning messy real-world knowledge into clean-cut mathematics.

---

business situations. To become a professional this is what you have to know. You must have proved yourself – academically – to be a fluent technician.

Why Technique?

The debate about 'scientific method' in OR has been predominantly about techniques. The debate about techniques is also, however, about the niche of OR in the modern commercial division of labour. OR workers do not work in labs. They investigate real systems which cause problems for managers, and they are employed within non-academic organisations to deliver results which make a practical difference to how the organisation runs day by day. By the early Sixties the situation was being mapped out, as in this quotation from an American book on mathematical economics:

> Businesses are employing individuals to apply OR methods to certain kinds of problems. On a novel or complex problem, it may be preferable to engage more than one person . . . Unfortunately the idea that OR is somehow synonymous with the team approach continues to be promulgated.

In a footnote the authors went on:

> There may be some vested interest involved here. Large consultant organisations which have such teams available may well be gratified by the emphasis on the team approach. But the businessman who is considering the possible application may well find the thought of a team of high-priced experts an overly chastening one.

And the conclusion: 'The businessman should be reassured that one man can perfectly well utilize OR methods.'

The authors are mapping the processes of political negotiation which take place between the world of The Organisation and the education industry. Consultants, especially in the US, carry a lot of weight and decline to act like employees. When the OR profession was becoming established and college-trained OR 'scientists' were entering the market, managers and businessmen needed to be re-assured that OR workers were just as employable as other kinds. These are not the eminent and promising academic researchers of wartime OR. They are college-trained technicians with no real research background, guardians of the toolbag, possessing specific narrow expertises which can be pinned on to specific, non-novel, non-complex problems designated by managers.

Skills and Certificates

This model of 'professionalisation' is not limited to OR. It was pioneered in the USA, during the fifty years before World War II by engineering professions which emerged in carefully graded hier-archies of skill, from drawing-board technique and routine clerical labour, up through detail design, to project engineering and project management. Each has its appropriate college-certified skill content and its professional police-system of institutions and standards boards. The model was established in engineering. OR came along after the war and was sucked in. Since then the same thing has happened to the computer 'professions'.

In the late Fifties computers were under the power of longhaired Wild Men, who ate and slept by 'their' machines, and who held the knowledge which could make them jump through hoops. It was a personal relationship. How could managers manage such idiosyn-cracy? In came the hardware and software firms, designing standard hardware, standard software, just as OR academics designed stan-dard stock-control calculation routines, and out went the hairies. In came the hierarchically graded, certified structure of a 'profession' consisting of coders, programmers, systems programmers, opera-tors, programmer-analysts, chief programmers, database managers. Each has a carefully graded entrance requirement, not because the skills are necessarily too complex to learn by 'sitting with Nelly', but because managers have wanted to reduce the actual complexity of work around computers until it is simple enough, in organisa-tional terms, to be overseen and controlled by managers who would otherwise be unable to monitor progress.

The Free Spirit of Knowledge

Skill reduced to formal qualifications, certifying a narrowly technical ability: the picture may be a little overdrawn. There are, for in-stance, some engineers who spend a lot of time thrashing out def-initions of problems and negotiating the worth of solutions, and who therefore relegate techniques and numerical methods to a sub-ordinate role somewhere in between. There are some OR scientists who are like this too, and some systems analysts. But not many. There is one fundamental reason for this, which is that not many workers would be allowed such a role in the hierarchical and generally conservative organisations which employ these skilled professionals. Most workers are not paid to think and certainly not to think too much. The engineers, systems analysts and OR scientists who stand out in this connection are people with the courage and the insight to wade into the political processes of The Organisation. Most professionals are happy to be technicians, and to succeed and be promoted even into managerial roles as technicians, protected from much of the flak by a shield of technique but imprisoned, by the same token, in a subordinate servicing role within a managerial hierarchy.

As scientists these professionals produce no generalised abstract knowledge. They produce particular, local and more or less concrete knowledges related to narrow, specific and more or less limited situations. In itself this is no bad thing, and no different from what most 'real' scientists do today. It is in itself no less noble to build a model of a spare parts warehouse in a car factory than it is to construct the growth sequence of the eggs of a nematode worm. There are differences, of course. The latter is part of an edifice of biological knowledge, of general interest across a whole range of countries and contexts, while the former is unlikely ever to see the light of day outside a particular company. This difference reflects the narrowness of commercial interests; most OR work is – contractually – unpublishable. But then, 'real' science is coming more and more under this restraint, especially in molecular genetics.

The difference between the two forms of knowledge also reflects real differences in generality between the kind of objects that managerial and biological sciences look at. Biologists' interest is more abstract and, paradoxically, this narrowness makes biological knowledge relevant to a geographically though not practically larger extent. These kinds of differences do not necessarily make the OR scientist, the engineer or the systems analyst inferior to other more stereotypical kinds of scientific worker as producers of knowledge.

Assumed differences between industrial 'professionals' and 'real scientists' become smaller and smaller on closer investigation. There is one profound reason for this, of course, and that is that most scientific research and development workers are industrial workers. The US National Science Board (*Science Indicators 1980*) estimated that out of 659,000 American scientists and engineers employed in research and development in 1980, 465,000 would be employed in industry. Universities and colleges trailed a long second with 82,500, though two-thirds of all scientific and technical articles emerged from this sector in 1979. The picture is similar in all industrialised countries.

But the similarities between industrial 'professionals' and 'real' scientists lie deeper, in the structure of science itself. If 'professionals' are essentially technicians, so are scientists generally. Researchers in 1670 were either professors (Newton, Malpighi), wealthy 'amateurs' (Boyle, Huygens) or ill-paid part-timers (Hooke, Flamsteed), and none of them *had* to do science. In 1770 the situation had not yet changed radically. But by 1870 science had become academicised, and in 1970 it would have been quite outrageous for anyone to presume to do research unless he or she had at least an undergraduate introduction to the chosen field. That is, by the end of the 1800s there were workers who knew no other work. They had been apprentice-bound over a period of years in a particular, narrowing, research area, and in each 'field' an established body of knowledge and technique dictated the norms of acceptable work. Entry to the professional community came through the subordination of individual ideas and interests to the corporate identity of the Field.

Science for the normal scientist is not breaking moulds and promulgating new over-arching principles. All researchers most of the time and most researchers all of the time are committed by their occupation to the modest and disciplined application of techniques and models. These are wielded in relatively narrow, specific and more or less limited situations defined in terms of accepted research objects and established lines of research. The kind of object studied in, say, biology differs from that studied by the OR scientist or the systems analyst, but the kind of work is the same. And like non-scientific workers, scientists are tied to their trade.

CRAFT AND CULTURE AND SCIENTIFIC WORK
Complex as it is, the division of labour within science cannot be understood on its own. Science needs to be understood and evaluated by locating it within the division of labour in society generally, as one kind of work among others. In valuing science we need to learn to look at the connections which exist between particular groups of scientific workers and other groups, including:
□workers in other specialist areas of science, who may not speak the same language or work in the same kind of economic institution, e.g., university, corporate lab, state research institute
□managers and administrators, and 'management scientists' who apply quantitative and logical techniques to provide managers with numerical answers to their questions concerning choice of action. These people tend to work under 'Management Services' cover rather than using a title containing the word 'science', even though their work is very similar in training and content to that of natural scientists
□data-processing workers, including clerical workers generally, and computer 'professionals' specifically
□manual workers of many kinds, both within labs (cleaners, technicians) and on the shop floor, where ideas produced in research and development are implemented via the mediation of managers

and more money

☐staff and students in educational establishments

☐designers of products and processes who use and commission and themselves carry out research investigations and, through them,

☐consumers of products and all workers whose work is designed for them (whose needs are anticipated, preproduced or pre-empted in the design and marketing of consumer goods and industrial equipment, from telephones to furnaces).

Some of these categories overlap. All their members, for example, are consumers of designed products. They all have connections in practice apart from those which operate through *scientific* work. They are related, for example, through schooling, through national and multinational firms, through government policies and political parties, through shared leisure and cultural interests, and so on. The connections are clearly very complex.

If this kind of picture were drawn in any detail it would soon be obvious that we can no more talk of 'the scientist' or scientific worker than we can 'the worker', and this for the same reason. The division of labour is too far-reaching, diverse and complex for single generalising terms to capture the content of all individuals' work. Yet there is a sense in which 'science' has a more concrete and complete meaning now than at any previous time in history. No longer a marginal activity dependent upon the wealth and largesse of a minority, research and the implementation of research-based decisions has become a quite fundamental cultural mode. It is now normal for most major changes in machines, tools, work methods, work organisation, products and processes to be based on detailed study and design by specialists outside the job, where only two hundred years ago such changes would have been made, albeit at a much lower level of intensity, on the basis of the accumulated experience and the craft traditions of the workers involved. The Industrial and Agricultural Revolution was the initial flexing of the 'modern' mode. The integrity of work and innovation has been and is being sundered and recomposed in a long revolution, along with the integrity of work and life, knowledge and practice. 'Craft' in work has been driven back since the late eighteenth century, as science in design has been advanced. For us, meeting this history now in the twentieth century, 'science' is one of the names for the whole totality of this mode of decomposing and recomposing society through work.

The Lost World?

One aspect of valuing this complex and contradictory whole is the problem of whether science itself is a craft practice and, as such, a repository of values that are alternatives to those of industrialised production. This is important because of claims that are continually made, reaching back to a supposed golden age in our grandparents' and great-grandparents' time, that science itself is a source of values such as progress, detachment, debate, intellectual honesty. The ethic is still quite active though less so than it was. But is it rooted in any coherent mode of practice, or is it just a detached rhetoric?

As an example of such rhetoric note how the distinguished Soviet physicist Peter Kapitsa, visiting England in the 1960s, lamented in sexist romantic fashion that

> The year that Rutherford died [Sir Ernest Rutherford, who 'split the atom', died in 1938] there disappeared for ever the happy days of free scientific work which gave us such delight in our youth. Science has lost her freedom. Science has become a productive force. She has become rich but she has become enslaved and a part of her is veiled in secrecy. I do not know whether Rutherford would continue nowadays to laugh and joke as he used to do.

This kind of romanticisation occurs in all fields: Kapitsa's retailing of a physics 'golden age' myth is parallel to 'pastoral' in literature. Not only is it parallel, but it also has the same history behind it.

The community of professionalised science in the imperialist era: a 'conversazione' in the Royal Society library at Burlington House, 1908. (Source: The Royal Society of London)

Capital moves in to organise and recompose leisure and work and, notably since Rutherford died and World War II began, scientists' work. Romantic reconstructions of bygone science are of little use in themselves because the freedoms were never as fully real as myth makes them. The freedom of classical science was freedom to publish but not freedom to do just any work, because research has always cost money and time, and the use of resources is always policed by social interests. Sciences' freedom is freedom to criticise within a corporate consensus, which leans the weight of seniority and patronage upon even unborn research programmes. These values, as idealised, are of course compromised in industrialised science. But they never were quite the values that they are in retrospect made out to be.

What we should read in such references to the past is not so much the past itself as the present: present day scientists' own sense of a necessary integrity, which they feel unable to honour fully in their present day work. It is the recognised need for intellectual integrity and the intimacy of 'craft' knowledge – as opposed to industrialised knowledge-manufacture – which needs to be taken into account in evaluating the places that scientific workers might occupy in changing the social relations of science. There are intellectual traditions in the sciences which still bear some of the marks of craft tradition.

Yet in another sense science has no tradition outside the present. There was no history, except a history of errors, before today's textbook and today's theory. Each generation rewrites the textbooks with a completeness which, if it were found in other areas of culture such as social history, would be called totalitarian. There is no future beyond Now – progress means just the same but more of it. This is the nature of a 'progressive' discipline. It is this more-of-the-sameness which has made science the envy of other academic pursuits, because it is the source of sciences' particular productivity. And it is narrowness, fragmentation and a systematic refusal to move outside the present, that is, to engage in real cultural debate, which are the conditions of this kind of progress. Whether it takes place within industry or not, science is an industrial mode. It is rigidly hierarchical, fragmented and routinised. The work of its (mostly lowly) practitioners is prescribed for them by the corporate logic of 'the field'. The ideal of science was industrial – Bacon, Petty, Babbage – even before there was any large scale 'scientific' industry, and normal science today is an industrial mode *par excellence*.

So what price the craft tradition of science, and the idea of science as a sphere of values alternative to those of an industrialised world? Was there ever a real craft tradition or was it 'always' (that is, since the mid nineteenth century, when science became a professional rather than an enthusiast's activity) an aggressive competition with nature and peers of the kind which, when we see it in present-day science, seems all of a piece with the rat race in business and commerce? Are the classical values of science mere rhetoric, or do they reside in practice, somewhere? These are not questions that can be answered in a small space, or simply, or at all in any adequate sense at the present. Too few people have tried to penetrate the mystique of science for there to be many reference points which we can take as firm if we want to understand sciences' relations with the world of work as we experience it.

It is certain that the traditional values of science are problematic, just as the status and relations of science as work are problematic. How do we balance objectivity and detachment against the context-bound narrowness of skilling that seems necessary to productivity in science? The 'body of knowledge'? How do we set this in relation to other kinds of knowledge like the arts, the humanities and our own personal knowledge of life, when the very fragmentation and technique-bound closedness of sciences defies attempts at integrating debate? The 'community' of science? How do we relate to scientific workers when they stand on their professional status and we are confused about the nature and organisation of *work*?

Why are Scientists?

FUNDS FOR INDUSTRIAL R&D IN DIFFERENT INDUSTRIES, UNITED STATES 1978

INDUSTRY	DOLLARS (millions)
electronics, communication	6,739
aircraft, missiles	7,700
machinery	4,469
chemicals and allied products	3,598
motor vehicles	3,913
scientific instruments	1,723
petroleum refining	1,071
primary metals extraction	546
food	428
fabrication of metals	397
rubber	495
paper	394
glass, stone, clay products	330
Total	33,400

Data: US Bureau of the Census, *Statistical Abstract of the United States 1980*, Washington DC, 1980, Table 1067 (Figures include federal and private funds.)
About 60% of scientists and engineers work in business and industry, 15% in education, 10% in federal government departments or agencies and 15% in other sectors.

SCIENTISTS, ENGINEERS AND TECHNOLOGISTS WORKING IN DIFFERENT SECTORS OF EMPLOYMENT, GREAT BRITAIN 1971

SECTOR OF EMPLOYMENT	NUMBER OF S/E/Ts	MAJOR SUB-SECTORS
agriculture, forestry, fishing	339	
mining, quarrying	391	(coal 291)
food, drink, tobacco	611	(brewing 111)
coal and petroleum products	364	(oil refining 315)
chemicals and allied products	2,767	
metal manufacturing	878	
mechanical engineering	2,078	
instrument engineering	422	
electrical engineering	2,997	(computers 687)
vehicles	1,452	(aerospace 910)
other metal goods	474	
textiles	437	(synthetic fibres 147)
leather, fur	17	
clothing, footwear	39	
bricks, pottery, glass, cement	380	(glass 155)
timber, furniture	43	
paper, printing, publishing	344	(paper and board 120)
other manufacturing	341	(rubber 144)
construction	1,726	
gas, electricity, water	1,714	
transport, communication	960	(post and telecommunications 310)
distribution	803	
insurance, banking, finance	1,112	(insurance 298)
professional and scientific services	13,092	(primary/secondary schools 3,757) (universities 2,025) (medical, dental 1,119) (architects, vets, etc. 1,878)
miscellaneous	679	(cinema, theatre, radio 197)
public administration, defence	3,114	(armed forces 510) (defence departments 370) (police 18)

Data: Department of Industry, *Census of Population 1971, Great Britain*, HMSO (Figures are numbers recorded in a ten per cent sample of the population.)

THE APPROXIMATE SPLIT OF WORLD R&D SPENDING is:

military and space research	33%
energy, health, pollution control, agriculture	25%
basic research	15%
others (including information processing, transportation)	27%

Who are Scientists?

Scientist and Engineer

Includes any person who has received scientific or technical training to the professional level (usually completion of a third level education) in any field of science; natural sciences, engineering, agricultural and medical sciences, and social sciences and humanities.

Technician

Includes any person who is considered qualified as such on the basis of having received specialised vocational or technical training in any branch of knowledge or technology of a specified standard (i.e., at least three years after the first cycle of second level education).

Research and Experimental Development

In general R&D is defined as any creative systematic activity undertaken to increase the stock of scientific knowledge and to devise new applications. It includes fundamental research, applied research in such fields as agriculture, medicine, industrial chemistry, etc., and experimental development work leading to new devices, products or processes.

Definitions: *UN Statistical Yearbook 1979–80*

NUMBERS OF SCIENTISTS, ENGINEERS AND TECHNOLOGISTS, GREAT BRITAIN 1971

	STUDENTS	EMPLOYED	UNEMPLOYED
Science agriculture, biology, chemistry, geology, mathematics, physics and general and other sciences (including earth sciences, forensic science, philosophy of science and meteorology)	**1,711** (1,381) (330)	**17,945** (14,883) (3,062)	**377** (271) (106)
Engineering chemical, civil and structural, electrical, mechanical, mining and other (including engineering design, industrial engineering, nuclear power engineering and jig and tool design)	**497** (492) (5)	**18,227** (18,157) (70)	**320** (318) (2)

(Table continued p. 75)

Technology
    metallurgy and other
    (including brewing,

building, ceramics, cloth	**86**	**1,842**	**33**
manufacture, food	(78)	(1,805)	(33)
science, leather	(8)	(37)	(–)

technology, paper
technology and timber
science)

Data: Department of Industry, *Census of Population 1971, Great Britain*,
HMSO (Figures as reported in a ten per cent sample of the population.)
In each column: figure in **bold type** is total, upper figure in brackets is
males, lower figure in brackets is females.
  Those employed are not necessary employed in research or development
work. For example, around seven thousand of those surveyed, out of
around forty thousand, were employed in teaching or insurance, banking
and finance. The survey noted workers' first qualifications and current
occupations, but the two are not necessarily connected.

UNITED STATES 1978
According to descriptions of their actual daily work, only 3 out of 10
scientists and engineers are doing R&D. Two out of 10 scientists (3 out of
10 engineers) are managers, and 4 out of 10 scientists (3 out of 10
engineers) do production-related work: reporting, statistics, computing,
production or inspection. One out of 10 do teaching, consulting, or some
other activity.

**MEN FULL-TIME WORKERS, COLLECTIVE PAY AGREEMENTS, GREAT
BRITAIN 1981**

AVERAGE GROSS HOURLY EARNINGS (Pence)

453	Professional and related in science, engineering, technology and similar fields [Post Office]
408	Police (below sergeants), public and private
361	Electricity power plant operators, switchboard attendants
336	Stevedores and dockers
318	Telephone fitters [Post Office]
304	Registered and enrolled nurses and midwives [NHS]
298	Locomotive drivers and motormen [British Rail]
287	Machine tool operators (not setting up)
260	Refuse collectors, dustmen
255	Carpenters and joiners
212	Caretakers; hospital porters
198	Agricultural machinery drivers/operators

Data: Department of Employment, *New Earnings Survey 1981*, HMSO,
1981

SALARIES, UNITED STATES 1978 ('000 DOLLARS)

engineering	29.5
health officers	40.0
nursing personnel	17.8
physics	31.5
chemistry	27.0
life sciences	23.9
mathematics	27.2
geography, cartography	23.3
psychology	29.0

Data: US Bureau of the Census, *Statistical Abstract of the United States 1980*, Washington DC, 1980, Table 1080.
In these categories, women's salaries are generally within 10% of men's, but lower.

MEDIAN WEEKLY EARNINGS OF FULL-TIME EMPLOYED MEN, UNITED STATES 1978

(average of all workers)	$227
professional – technical	$294
clerical	$175
craft workers	$279
operatives	$191
farmworkers	$139
private household workers	$ 59

Data: US Bureau of the Census, *Statistical Abstract of the United States 1980*, Washington DC, 1980
The 'median' is the middle value in the whole range of earnings. It is not necessarily the same as the arithmetic average.

QUALIFICATIONS RECOGNISED IN CENSUS OF SCIENTISTS, ENGINEERS AND TECHNOLOGISTS, GREAT BRITAIN 1971

Here are some samples of the qualifications recognised in the 1971 Great Britain Census of Population. They were sorted into five groups which, although the Census report does not say so, were clearly supposed to reflect some kind of intellectual peck order in the minds of the evaluators.

Group 1
University degrees (Batchelors' at most universities, Master of Arts at Oxbridge) including Colleges of Advanced Technology (CATs)

Group 2
Degrees awarded by the Council for National Academic Awards (CNAA): polytechnics award degrees through this system

Group 3
Diplomas of Technology, awarded by polytechnics and CATs

Group 4
Associateships, diplomas and similar qualifications: e.g., Associate of the Bradford Institute of Technology, Associate of the Royal Veterinary College, Diploma of the Birmingham College of Art in textiles

Group 5
Membership grades of professional institutions: e.g., Associate of the Institute of Physics, Fellow of the Society of Dyers and Colourists, Graduate of the Institute of Production Engineers, Graduate Member of the Institute of Electronic and Radio Engineers

Grouping these together, the Census gave the following figures (from a ten per cent sample of the population in 1971):

TOTAL	38,744
graduates	25,931
associates	706
professional members	12,107

Overlap between the categories was removed by putting members of the 'lower' category (e.g., members of professional institutes) into the appropriate 'higher' category (e.g., graduates). Thus the 12,107 members of professional institutions are non-graduates.

Where are Scientists?

DISTRIBUTION OF SCIENTISTS, ENGINEERS AND TECHNOLOGISTS IN THE UK, 1971

REGION	ECONOMICALLY ACTIVE (UNEMPLOYED)
Yorkshire and Humberside	2,416 (51)
Northern Region	1,829 (38)
North West Region	4,561 (86)
East Midlands	2,074 (37)
West Midlands	3,189 (48)
East Anglia	1,055 (17)
South East	16,298 (311)
South West	2,546 (69)
Wales	1,495 (31)
Scotland	3,281 (42)

Data: Department of Industry, *Census of Population 1971: Great Britain*, HMSO (Figures are the numbers recorded in a ten per cent sample of the population.)

DISTRIBUTION OF INDUSTRIAL R & D FUNDS IN THE US, 1978

DIVISION	% of TOTAL POPULATION	% of R&D FUNDS
New England	6	8
Middle Atlantic	17	21
East, North Central	19	21
West, North Central	8	5
South Atlantic	16	7
East, South Central	6	2
West, South Central	10	4
Mountain	5	3
Pacific	14	21

Data: US Bureau of the Census, *Statistical Abstracts of the United States 1980*, Washington DC, 1980

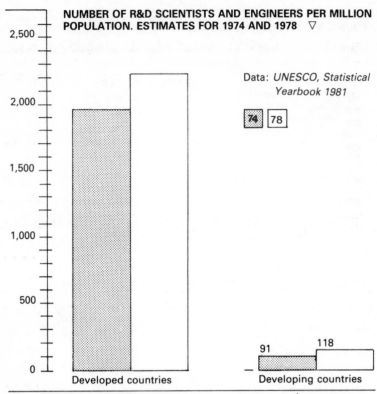

NUMBER OF R&D SCIENTISTS AND ENGINEERS PER MILLION POPULATION. ESTIMATES FOR 1974 AND 1978 ▽

Data: *UNESCO, Statistical Yearbook 1981*

74 78

2,500	
2,000	
1,500	
1,000	
500	
0	

Developed countries

91 118

Developing countries

DISTRIBUTION OF R&D PERSONNEL AND EXPENDITURE BY REGION. ESTIMATED PERCENTAGES 1978.

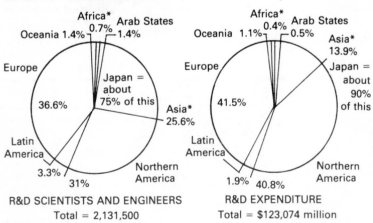

Africa* 0.7% Arab States 1.4%
Oceania 1.4%
Europe 36.6%
Japan = about 75% of this
Asia* 25.6%
Latin America 3.3%
Northern America 31%

R&D SCIENTISTS AND ENGINEERS
Total = 2,131,500

Africa* 0.4% Arab States 0.5%
Oceania 1.1%
Asia* 13.9%
Europe 41.5%
Japan = about 90% of this
Latin America 1.9%
Northern America 40.8%

R&D EXPENDITURE
Total = $123,074 million

*Excluding Arab States   Data: *UNESCO, Statistical Yearbook 1981*

▽ *Not including the USSR, China, Mongolia, Vietnam and the Democratic People's Republic of Korea, for which country comparable data for 1974 and 1978 are not available.*

SCIENTISTS, ENGINEERS AND TECHNICIANS DOING R&D IN SELECTED COUNTRIES IN THE 1970s

COUNTRY	YEAR	SCIENTISTS & ENGINEERS	TECHNICIANS	TOTAL, PER MILLION POPULATION
India	1971	28,233	25,872	99
United Kingdom	1971	79,300	75,800	2,794
United States	1970	560,800	44,800	2,983
Federal Republic of Germany	1970	103,857	100,276	3,319
Japan	1975	407,192	87,783	4,423
USSR	1979	1,279,600	–	4,876

Data: *UN Statistical Yearbook 1979–80*

USSR figures include workers in higher education. US scientists and engineers figure includes higher education, technicians figure is higher education only. Other figures exclude higher education.

Which Are Scientists?

DEPT. No. _167115_   Gd. _AP 21B_

d.o.b. _12.1.45 (37)_   S. _M_

Post _Research - Unit B716._

Degrees _BSc. (Hons) - 2.1 (th) University of_
Manchester - Inst of Tech.

The managerial representation is of scientists as 'manpower' — packages of potential, documented by achievement in formally certifiable ways and assessed objectively in matching social production (of masses of workers) with social needs
(the needs of production). Above are the facts of scientific workers' official
histories.

    As personal histories, however, many fictions are created and used in living
the facts — both self-images which give the work meaning to workers, and images
for public consumption in fiction.

All of these work within the division of scientific and
technical labour. But which would normally
be dignified by the description
'scientist'? And who decides
who gets the status?

student

woman packer

engineer

research manager

post-doctoral researcher

health and safety inspector

Nobel prize winner

school teacher

process plant operator

bottle washer

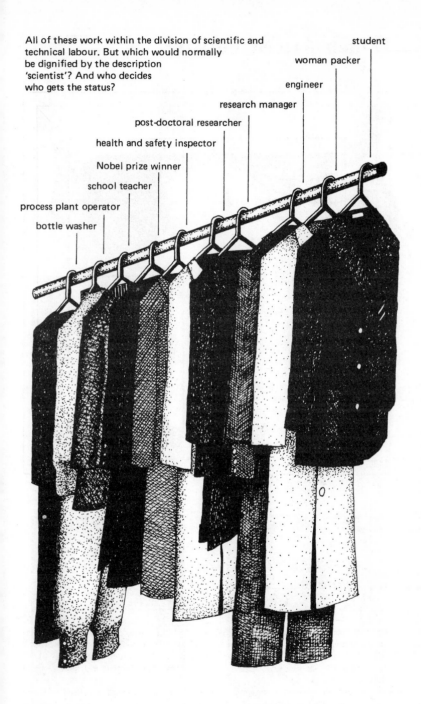

FACTS OR FICTIONS?

1. explorer
2. sage
3. altruist
4. tragic hero
5. researcher
6. mystery man
7. judge
8. nut case
9. whizz kid
10. arch-villain
11. proletarian scientist

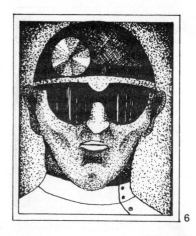

6

7

8

9

10

11

5   Uses of Science

SCIENCE AS AN IDEAL: DOMINATION OVER NATURE

Science before it became science was and still is ideology. The earliest and perhaps still the most profoundly effective social role of developed science was as a projection of a specific kind of social interest. Writing during the sixteenth century, Francis Bacon prophesied a new empire of knowledge and these prophesies were taken up as a battle call by natural philosophers in the century after Bacon's death. In earlier centuries, whatever the differences that were imagined in the actual composition of matter (the Four Elements, the Four Humours, and so on), views of nature had regarded values as being intrinsic in the material order of the cosmos. Certain

HEADPIECE The image of pure science: Albert Einstein fancy free at a friend's house near Los Angeles, 1933. (Photograph: Archives of California Institute of Technology)

forms, into which matter might be shaped, had intrinsic virtues and statuses; earthly matter, for example, had an intrinsic tendency to move downwards. The 'final causes' – which would today be regarded as the external purposes – of transformations were intrinsic and hierarchically organised within a system of higher and lower forms.

During the seventeenth century this kind of immanent value began to be displaced from views of nature so that, gradually, the modern sense of nature as 'mere' matter emerged as dominant in some circles. Values then were understood as external to nature, inherent in human and Godly purpose. On this understanding natural events could be studied neutrally without looking for hidden purposes in nature, and power over nature became a meaningful way of justifying and ordering the pursuit of knowledge.

The intellectual change was deep, and deeply tied to changes in the class order of society. A more active view of nature was the 'natural' outlook of rising classes in seventeenth-century society: merchants, 'improving' landlords, administrators, and others with an interest in the development of manufactures and organisations. In the century of the Puritan revolution in England, when the growing power of these classes was asserted against that of court-favoured magnates and traditionalist aristocratic landlords, science became part of the cultural identity of those who endorsed and propagated mobile and calculating anti-traditionalist values, the quickening spur of honour and preferment 'open to desert'. Religion and politics were ruled out of order in the world's first scientific society, the Royal Society, not because of the 'neutrality' of the merchants, artisans and others who supported the Baconian creed of material power, but because it was politic to suppress views on such matters at a time when power broking was a very active process. Key members of the Royal Society were 'Latitudinarians', supporters of Cromwell during the Commonwealth but conformers after the Restoration. Some radical Puritans even succeeded in transforming themselves into Church of England bishops. Thomas Sprat, biographer of the Royal Society, was one; John Wilkins, Cromwell's brother-in-law and Secretary of the Society, was another.

The proclaimed neutrality of the Royal Society in the pursuit of effective material power, and its charter as the *Royal* Society for Improving of Natural Knowledge, are aspects of the great English compromise between the new commercial classes and the established aristocratic order. The new philosophers, Isaac Newton among them, settled into the new order of accommodation between merchant and landed capital. Although it specifically concerned the production of knowledge, the 'New Philosophy' rose as a model of ordered, disciplined, hierarchical *production*. The Royal Society failed to bring about the Baconian project of radically transformed technology on any significant scale. But the ideological projection of a devalued, merely-material, manipulable Nature, subordinate to a purposeful and hierarchically ordered social machine, went very deep indeed in the culture of an emergent capitalist order.

RESEARCH AS DEVELOPMENT
It comes as something of a shock to recognise that one of the central images of industrial revolution in Britain was generated without the blessing of science. The dark satanic mills and their rows of carding,

Early English carding machine

spinning and weaving frames, driven first by water and then by steam, were in no strong sense a product of scientific research. Hargreaves' spinning jenny for weft threads and Arkwright's water frame for warp were under development from the 1760s, so that by 1780 there were twenty thousand of Hargreaves' machines in operation and the first water-driven carding and spinning mill was running in Lancashire. Yet these engines of revolution were developed without the kind of applied general knowledge and systematic uncovering of new knowledge that we refer to casually today when we speak, as if in a single word, of '*research* and development'.

Science was not invisible during the last half of the 1700s. James Watt's refinement of the steam engine involved new insights into the theory of heat, such as the ideas of specific heat and latent heat. Watt was brought into contact with the technology of steam at Glasgow University where he was a technician who repaired scientific instruments. Experimenters at the university had close links with the Carron ironworks and with other local industries. Yet in this case, as in most others, practical developments were ahead of theoretical insight. In agriculture and in other traditional sectors of production such as brewing, which had become a large scale industry well before the Industrial Revolution, many researches emerged out of technique. From the mid seventeenth century, experimental-

88

ists and natural philosophers began to take cues from industrial practice and tried to feed their discoveries back into industry. Often, however, they had more effect as disseminators of technique than as designers, and up to the middle of the nineteenth century technique, broadly speaking, led science.

Nevertheless around 1800 it is possible to see the relations between scientific investigations and technical practices beginning to shift beyond the connection in which brewing prompted chemistry and pumping engines prompted thermodynamics. Experimentation became established as a cultural mode in eighteenth-century middle-class life so that the liveliness of thinking in science and its actual models of natural order began to inspire and inform thinking in other spheres. But it also began to make its mark on the way in which technical investigations were conducted, and so to usher in a new connection between science and technique.

Research and Industry

Science as method began to infiltrate the practical world of technical development. Josiah Wedgwood of pottery fame was an active member of the Lunar Society, a scientific society based in Birmingham, and a profuse correspondent on matters of scientific knowledge. He filled his 'Commonplace Book' with minutely quantitative notations of everything which interested him and there are pages upon pages of careful calculations on, for example, the threads per inch of Lawns used for sieving clay. His systematic experimentation which led to new forms of china has become legendary, with trial pieces of the famous jasper running into tens of thousands. Certainly, an entrepreneur who would market black basalt china to exploit a current female fashion for bleached hands, and who would respond to the Flour Tax of the Napoleonic wars by producing piecrust ware, was not likely to miss a commercial opportunity provided by a new experimental technique or scientific theory. Wedgwood's directorship and personal involvement in development research gave it a highly systematic and quantitative cast. At the turn of the eighteenth century science at the hands of cultured entrepreneurs like Wedgwood began to take on its characteristically modern form as an input to commercial innovation, a model of method and also a system of ideas which could be embodied in new processes and products.

The systematic application of research effort to problems relevant to commercial practice took off most notably in Germany, where the chemistry teaching laboratory of Justus von Liebig (founded at Giessen in 1824) became a model of future development. During the 1830s universities in Germany vied with each other in establishing scientific chairs and by the 1840s results in nutrition (especially relevant to military needs) and agricultural chemistry had begun to have a wide impact on practice. Throughout the nineteenth century research establishments which were based on the German model

Josiah Wedgwood, Fellow of the Royal Society and the Society of Arts, b. 1730, d. 1795, 'who converted a rude and inconsiderable manufactory into a elegant Art and an important part of National Commerce'. (Source: Monument, parish church, Stoke-upon-Trent)

multiplied in universities, in private companies and eventually in state research institutes. By 1900 the six biggest German chemical works, which were closely connected with the steel industry because of its coal tar by-products, had more than six hundred and fifty chemists and engineers. Britain and the USA had around thirty or forty between them.

'Technology' in the modern sense of science and technology emerged during the latter part of the nineteenth century. In some cases traditional industries like chemical manufacturing became

90

'scientific' with the creation of quality control and research laboratories. In other cases large industrial firms moved in on science to serve their existing commercial needs and to exploit new markets. General Electric (created specifically to control patents), American Telephone and Telegraph (AT&T) and DuPont pioneered industrial research laboratories in the USA during the 1890s. Small firms depended for their scientific input on state-supported laboratories such as the National Physical Laboratory (Britain) or the National Bureau of Standards (USA) and, in America, also on private contractors such as the Arthur D. Little corporation.

Gun Cotton Diplomacy

Research-based technique had possibilities which were exploited differently in different political-economic contexts, and appeared in various strategies adopted during the capitalist accumulation crisis of the Great Depression, 1873–96. Two broad approaches were adopted; opening up new markets and territories for investment – that is, imperialism – and more intensive exploitation of labour within industry. Richer nations such as Britain took the former road, opening a great storehouse of materials for cataloguing and investigation by researchers, especially biologists, geologists and doctors. Imperialism set many technical challenges in conquering terrains and environments that were physiologically, culturally and militarily inhospitable.

To an extent The Frontier functioned as America's internal empire, but Yankee ingenuity was directed more towards the second strategy, in the form of mechanisation. After World War I Germany, as a debtor country already industrialised, necessarily had to go down the second path of 'rationalisation', followed reluctantly by the weakened victors of the European war. Britain delayed modernising and the USA, now combining both strategies, became the dominant capitalist power. In Italy, a country with regions which even today are obscenely poor, Antonio Gramsci, the communist organiser and Marxist thinker, spoke of 'Americanism and Fordism' and the Turin auto workers knew what he was talking about. Economic development had become world development and the exploitation of nature was world-wide exploitation. The labour of science was worldwide too, not just in the interchangeability of its plug-in knowledge (an aspect of an international division of labour in science) but also in the ways that it was central in internal and external strategies of economic expansion.

By the end of the nineteenth century the detailed division of labour and the systematic development of research in chemistry had produced a body of natural knowledge and investigative technique which often allowed manufacturers in the industrialised countries an effective choice, between extraction of substances from imported materials and synthesis from coal tar derivatives. A British chemist in 1931 (in *Chemistry Triumphant*, part of a series called 'Century of Progress') gave an account of successes in the dye industry. In Germany in particular, imperialist competition prompted new leaps in synthetic chemistry. In 1909 Friedlaender, working on varieties of indigo which is the dye now used to make blue jeans blue, synthesised a product identical with Tyrian purple. Previously this royal dye of antiquity had required twelve thousand Mediterranean murex shells to yield less than two grams. After 1909 the market for the traditional product ceased to exist. Many other natural products had been similarly substituted. The need for natural cochineal (used as a food dye) vanished, as did the livelihoods of thousands of Brazilians who caught and processed the cochineal insects. Quinine was synthesised and the basis of Peru's economy crumbled almost overnight. Phosphate chemistry did the same for the guano industry of Chile, substituting synthetic fertilisers for the faeces of sea birds.

Chemistry had a distinct political use value. Management in the Lancashire cotton industry, for example, required a careful sense of imperialist diplomacy. In the colonies it could be possible to destroy a British-managed indigenous extractive industry *and* a captive market for British finished goods by choosing a synthetic route for dyestuffs. The productivity of detail labour in research (which is what the 'century of progress' signifies) meant according to *Chemistry Triumphant* that, 'Discovery is now far ahead of practice. We now know the constitution of many naturally occurring compounds for which we have no adequately serviceable manufacturing steps.'

COLONIAL PREMIERS: "Well, good-bye, John, we have thoroughly enjoyed ourselves, We are proud of our Queen, and of our Empire, May the sun never set upon it."
JOHN BULL: "It's not likely to."
COLONIAL PREMIERS: "Never, so long as we get such a right royal SUNLIGHT welcome, SUNLIGHT weather, and plenty of—er—"
JOHN BULL (Smiling): "SUNLIGHT SOAP."

This advertisement for Sunlight Soap anticipates a continuing history of profitable metropolitan relations with countries on the imperialist periphery for Lever Bros, now Unilever, a food and chemicals multinational. (Illustration: *Illustrated London News,* Aug. 28, 1897, p.301.)

This turnaround in relations between science and technique, this cornucopia of science, had a dual aspect for the peripheral nations of imperialism. It was a Pandora's box that released in the colonies the stinging, biting evils which civilisation carefully sifted out from the 'neutral' clay of scientific knowledge, with which metropolitan industries built their magnificent edifices. It was also a sword of Damocles, hanging over the heads of colonial populations, maintaining their dependence yet threatening to cut off their livelihoods by turning laboratory knowledge into manufacturing technique if political situations got out of imperialist control.

Gun cotton, nitroglycerine, and the whole armoury of modern firepower dates from this period of scientific progress. The fortune which finances the Nobel Prizes came out of marketing dynamite, a stable, safe-to-handle form of nitroglycerine. The long established principles of ballistics were not very effective before the imperialist era because guns and projectiles and propellants were not of uniform quality. It was not until around the time of the Crimean War that chemical and metallurgical knowledge made accurate gunnery possible. Imperialist chemistry is an aspect of gunboat diplomacy not just because of this connection, as a supplier of military and commercial explosives. Systematic and redundant knowledge ('pure' science) also gave imperialist powers increased flexibility and leverage in their political-economic strategies. The community of science received shameful recognition of this in the name sometimes given to the 1914–18 war – 'The Chemists' War'.

SCIENCE AS CULTURAL RECOMPOSITION
In their second major use the natural sciences were agents of industrial and political recomposition. As this became established it prepared the ground for a third use which meshes closely with it: science as a basis of 'cultural' recomposition. In the form of machinery, and especially the machinery of mass production, science is a lynchpin of the domestic strategies of imperialism which are aimed at transforming the culture of work, in order to bring about more intense and fully controlled exploitation of 'internal' resources of labour in metropolitan countries. But science also figures in a domestic strategy complementary to this, that of increasing consumption in parallel with production so that science appears on the other side of the equation of mass production. In the same way that scientists originated 'time and motion study' techniques which were used in factories from the late nineteenth century as a means of increasing production, science in the hands of home-orientated experts became 'domestic science'. The washing machine and the sewing machine made consumer skills out of domestic chores, which experts could pronounce on. At a later stage they were transformed by the domestic electric motor, and other appliances followed. And just as it was a managerial principle that workers should be told by

science (managers' science) how best to produce, so it became a marketing principle that end users, defined now as 'consumers', should be educated through research to become efficient consumers. Advertising became a major industry.

Gathering momentum in America during the economic crisis of the late 1800s, these new kinds of connections between scientists, now taking on the modern public-relations role of 'experts', and working people, spread with the techniques of mass production. They became an integral part of the intervention by capital into cultural life at work and at home which characterises modern in-dustrialised societies.

An illustration from the 1980s is perhaps best here because divisions between work and home are currently shifting and, in places, breaking down as the machinery of production merges with the machinery of domestic consumption. This can be seen in the 'sa-tellites' sector of technical innovation – the apparatus of information transmission and processing.

All Keyed Up: Viewdata

Many of the key developments concern cable transmission, and satellites are the international link in the cable chainwork. The cable in question is not the kind of thing that we have lived with for a century now – metallic, heavy, bulky, carrying electrical energy at high or low power levels between generators and motors, between transducers and amplifiers, between telephones, between reception stations and 'piped' TV receivers. The cable referred to in current discussions of 'cable TV' is relatively lightweight. It is made of glass fibres. And it does not carry electricity but light – laser light, a very regular, concentrated and controllable form of light energy. Pulses of laser light can carry coded information very fast and efficiently, and a single optical fibre or 'light pipe' can carry a much denser and more mixed flow of information than an electricity 'pipe', that is, a metal wire.

In Belgium cable connections have been installed in nine-tenths of the homes and the German post office intends to have connections to all homes by the 1990s; Germans pay involuntarily for this 'home improvement' through higher rental charges. British Telecom is a long way behind the field, but it will be obvious when they start because pavements from Southampton to Aberdeen will be torn up to lay the cable. By the mid 1990s twelve per cent of the British Telecom network will be optical fibre and another thirty-six per cent will carry coded pulses by other means. The motivation for telephone businesses is that markets for television sets are now saturated and new forms of 'information' are being developed as commodities, as a springboard for renewed capital accumulation. Viewdata is one of these forms, which enables subscribers to dial 'pages' of data to be displayed on TV screens. Controllers of telephone networks expect to be able to take a slice of all the action by charging a fee for every item of information transmitted in addition to the income from equipment rental and maintenance, while at the same time entrenching further their positions as brokers of the means of communication. The German post office hopes to use the cable network to carry the existing TV channels as well as viewdata and similar new commodity services.

Access to Information

Britain's viewdata service (called Prestel) has been based on the principle that British Telecom is a common carrier, imposing no censorship on the matter transmitted. But at a government-sponsored conference early in 1982 it began to seem that this principle might disappear. Television set manufacturers, microchip firms and data-selling firms agreed that it was feasible to aim for a mass viewdata market by 1984, and are pressing for a simplified price structure for users of the service which would favour big commercial information-sellers over small and public-service suppliers of information. 'Technical' considerations go this way too. A proposed consumer package outlines changes in the format of data, cross-referencing between pages, and policing of the up-to-date status of information in order to increase the use value of the system to casual and inexpert users. The intention is to make it more 'user friendly', as computer professionals say. This sounds fine, but the labour involved in such editorial control would mean in practice that decisions would be taken concerning the 'marketability' of different kinds of information. Censorship.

Viewdata is essentially a computer system as are modern telephone networks, such as the German national system which automatically records all details of telephone calls and is therefore also a computer-based surveillance system. Software has to be written to control the way in which computers process data, so that certain kinds of input like enquiries, information, or directives, will be acceptable from users and certain – necessarily limited – options of use are offered such as communication link-ups, data from data banks, and processing of data to produce reports. This means that the workers who write software have to pre-think the uses that the computer will be competent to support. This is called preconceptualisation. All the user can do is to use or not use or sabotage the options provided.

This systemic and absolutely necessary narrowing of the potential variety of uses takes place in the design of any hardware or software system. But preconceptualisation is an especially dramatic and visible relationship between experts and society in computing, because it is a central cultural resource whose variety is being limited. Information is vital to democratic action, and becomes more so as social and technical systems become more complex and extensive.

Preconceptualisation – or, Who Makes the Rules – is a central issue of technology generally. But viewdata brings 'technical' issues right to the centre stage of cultural debate. American market forecasts project twenty million viewdata sets in home use by 1990. European forecasts are similar. National viewdata systems are being linked by satellite communication. What are the options? Who is making the rules of this game? Whose world will tomorrow's be? When we buy the new addition to our ideal home in 1984, which social interests will be getting the use value? Will the dominant use value be investors' profit?

Shifting the Middle Man's Burden
Sellers of services such as banking and retailing are very interested in cable computer systems, hence the pop image of 'the cashless society'. With viewdata – and a lot of non-altruistic investment and software development – you could pay in and draw out money from the bank, book holidays from home, do shopping from your arm-chair. If the State gets in on the act you might even complete your tax return by keying data directly into an Inland Revenue computer file. When such developments are implemented the seller of the commodity, say, a rail ticket, is shifting some of the necessary labour from their establishment on to you. While it seems to you that you are increasing your free time since you can carry out many transactions without travelling, you appear to the owners of the travel agency as an unpaid casual worker. Once they have paid for the labour of the software writer they can get rid of counter clerks. This is the domestic analogue of 'electronic mail' in the business world; by enabling more senior staff to communicate directly via a data network such systems do away with low-grade clerical jobs.

If this happens it will be a great leap forward in a process which has been going on since the time of Henry Ford. For up to now the car has been central in shifting labour from the sphere of paid to unpaid work. Store-bought goods used to be delivered but the supermarket shopper needs a car to take advantage of the 'conven-ience' and the price cuts, and thus becomes an unpaid replacement for a worker in the retail sector. In the main it is women, as domestic labourers, who have taken over the tasks shifted in this way, house-wives replacing delivery boys, one kind of cheap (actually, unpaid) service labour replacing another. The home computer terminal, linked through cable and TV screen to viewdata computers and data banks, opens up great new vistas of labour-shifting into the home, new vistas of cost-cutting for capital accumulation.

At the same time 'shopping by computer' is a prime case of capitalist socialisation of labour. The information-processing system links workers in the home and in employers' workplaces over a wider geographical range, making possible a widening range of transactions between them. The apparatus of connections and data

CONMAN DESIGN ASSOCIATES
PLAN FOR 44 ACACIA AVE
© 1982

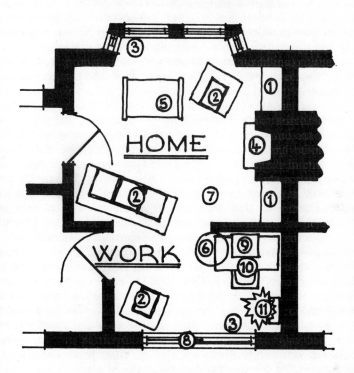

HOME

WORK

1. Reader's Digest 1956-79
2. G-Plan suite 1969
3. Dralon velvet curtains (pampas) 1973
4. Log-effect gas fire 1975
5. Habitat pine coffee table 1975
6. Habitat standard lamp 1976
7. Acrylic berber carpet 1978
8. Slik-slide double-glazed
   aluminium patio door 1979
9. Sony Trinitron TV 1981
10. Olivetti DE 700 workstation 1982
11. clocking-on machine 1984

gets bigger and bigger. But the computer shopper doesn't meet the warehouse clerks, the despatchers and the postal delivery workers, in the way that they would have met and maybe talked to or asked advice from the saleswoman or man in a shop. The process is 'social' only in a reduced technical sense. Objective – not subjective – socialisation of labour is what takes place when capital investment is in command. In a shop the shopper might have taken a cue from the appearance and location of the shop or the people in it or the things they said or the way they spoke. Isolated at home with the glossy images of the retailer's catalogue there are none of these cultural cues to go by, and the few new ones are different and harder to read. The shopper is trapped in the ideal home, confronted by the commodity spectacle but without awkward distractions in the form of other human interests. Meanwhile, behind the screen, the much-vaunted free time generated by automated systems appears in its two characteristic capitalist forms; free time for the individual as a buyer of commodities within the confines of the family economy, and time 'saved' by sellers of commodities which is what we register, at a social level, as unemployment and underemployment.

One more turn of the screw. Because many jobs – clerical, administrative, technical – revolve around the processing of data it is possible, through computer terminals installed at home, to decentralise significant amounts of the work now normally done in employers' workplaces. As professional and non-professional work is transferred to the home, and as the computer keyboard becomes more common in workplaces generally, workers' habituation to the keystroke form of communication will enable them to slip more easily into new habits of consumer living. With the keyboard workstation installed at home the dividing line between work and leisure may start to get blurred. Having a workstation at home will make extra overtime more tempting as an alternative to helping the children with homework, or just chatting. Underlying the limited autonomy which the individual worker-consumer gains in this way is the larger and dominant autonomy of commodity-sellers in relation to a fragmented 'community', a society whose socialisation is capital and profit. Profit decides if it's progress.

SCIENCE AS SCHOOLING
Experts in the sphere of culture have not so far had the same force as experts in the culture of work. This must be due in large part to wage work making workers subject to managers' prerogatives in ways that private consumption does not. Nevertheless, domestic and personal experts like home economists, consumer goods designers, psychologists, doctors and psychiatrists, do carry real influence. Much of it is based in the institutional power of a fourth, long established, use of science: in education.

The early history of science in education is fascinating because it is so different from the education with which we ourselves have grown up. Formal education today is compulsory and universal, and still centres on the Victorian Three-R's. Science, though politicians endorse its importance in the same way that they used to kiss babies, is marginal in many a curriculum. Contrast this with the late eighteenth century. Science then was unimportant, judged by traditionally established patterns of education for the ruling classes in universities. The facts which mattered were still the 'facts' (syllogisms and so on) of the old scholastic disciplines and the classics – the same static monopoly of knowledge against which the 'new philosophy' of observation and experiment had set itself in the seventeenth century. Education elsewhere – in France, Germany and Scotland – had been more responsive to the empirical and exploratory interests of the new philosophy. But in England the conservatism of the established Church allied with traditional definitions of breeding and culture placed science beyond the pale, along with the non-aristocratic classes whose interests in science were the strongest. Although capitalism became a real force during the eighteenth century, and became dominant in many sectors, it did so under the cloak of a continuation of the traditional order. Aristocratic landowners gave way to, or became, agricultural capitalists, but the social forms as constituted by the landed estate and the landlord kept culture on a rein.

To the extent that it was a part of formal education in England at all, science during the late eighteenth century was part of a radical and alternative political culture, an education self-organised by religious radicals because they were excluded from the mainstream institutions of their day by the Test Acts (which 'tested' religious orthodoxy). Many captains of manufacturing industry were schooled in the 'Dissenting Academies' of the methodist sects. The use of science within this subculture was significant, for it not only represented the intended means of material control over the forces of nature – the Baconian project of the New Philosophy – but also symbolised a challenge to the traditional social order, from marginalised but increasingly coherent interests. This marginal culture included a strong element of what was stigmatised as 'Jacobinism' – people who were in sympathy with what became the French Revolution. Such was the cultural location of science during the early years of the Industrial Revolution. Joseph Priestley, the chemist now famous for his researches on gases, was a Unitarian minister, a lecturer and educationist connected with a number of dissenting academies and was offered, along with the revolutionary pamphleteer, Tom Paine, a seat in the French post-revolutionary Convention in 1792. France was the main centre of chemical researches at that time, and Priestley was a respected figure.

It was not surprising that the radical alliance of science disappeared as a result of the force with which Jacobins were repressed in the final decade of the century. Priestley's Unitarian meeting house and his own home were fired by an anti-Jacobin mob in 1791 and he felt it necessary to emigrate to America in 1793. James Watt, a member of the Lunar Society along with Priestley, Wedgwood and Erasmus Darwin (Charles Darwin's grandfather), is said to have attended a meeting following the 1791 riots with pistols in his pockets. When the manufacturing classes became established politically, through the reform legislation of 1832 and the repeal of the Corn Laws in 1842, it was not as political radicals in the revolutionary mould. Nor indeed was it as prophets of science. University reform began to move the establishment some way towards radicals' intellectual commitments: University College London was founded in 1828, with scientific topics among the main teaching subjects. Conversely, political reform had begun to move the cultural commitments of industrial capitalists towards more conventional definitions and institutions. Tory Prime Ministers and Herefordshire gentry emerged two generations on, in families that had supported Priestley's science and politics. Their children were primed for Oxbridge, not the new 'redbrick' colleges. No more dissent – the English Compromise again.

The big bourgeoisie abandoned science as a cultural pursuit for their own advancement during the first third of the nineteenth century. This withdrawal was accompanied by a new identification of the uses of science in educating lower classes. Workers' own educational institutions, such as 'Corresponding Societies' which had Jacobin sympathies, were systematically suppressed during the 1790s, and subsequent initiatives like Mechanics' Institutes (the London Institute was founded in 1823) were carefully monitored and policed by middle-class board members. Although ultra-conservatives thought that any education of the lower classes was dangerous, others felt it was necessary to allay the unrest becoming apparent in industrial cities. Mechanics, skilled operatives and artisans were literate people who, if undirected, might read the wrong things. The Institutes offered them the 'pure' but factual knowledge of physics, chemistry, mathematics, earth sciences, biology. Phrenology, as an improving philosophy and science of mind, was popular. As the artisans measured each others' craniums they could speculate on their chances within a competitive and individualist social order. This curriculum, it was suggested, 'reclaimed many from the habits of vice. It provided them with safe and rational recreation . . . and had the effect of promoting the strength and prosperity of the country in general.' By the mid-1820s the project of modern mass education had been identified and science, as the model of down-to-earth rationality, discipline and moderation, was at its centre.

By the second generation Mechanics' Institutes had been taken over by the lower middle classes. Left behind by the elevation of industrialists into traditional ruling class culture, these middling people adopted the hand-me-down culture of self-improvement through science as their own pathway to moderate success within an increasingly technical system of production and administration. A more highly industrialised pattern had been set for the education of working-class children. The Lancastrian School in Manchester is one example, which in 1809 processed a thousand children under one roof, reciting their texts in unison under the supervision of child monitors. Science became middle-class property. The workers had to make do with Reading, Writing and 'Rithmetic. The middle classes could learn to think – within moderation. The children of the 'labouring poor' would learn to labour.

SCIENCE AS MANAGEMENT

By the beginning of the present century many of the fundamental patterns in the uses of science had been set – in ideology, in technical development, in the organisation of public and private life and in schooling the population of industrial societies. Though many large and small changes have taken place in these patterns during this century, one significant addition to the repertoire which has emerged over the past fifty years is science as a form of management.

New management methods have been demanded since World War II not only by the continued expansion of state administration and business but also by massive scientific projects like atomic weapons research, nuclear power, space exploration and weapons systems development. Some sectors of 'pure' science (high energy physics, astronomy and oceanography are examples) have extended their apparatus on such a scale that it requires a considerable management effort to service it. In this general context there has been a self-conscious application of forms of organisation and analysis taken from scientific work, which have resulted in institutions such as the 'think tank' and conceptual approaches such as 'systems thinking'. Science has been drawn beyond the design of particular pieces of equipment towards the design of policies, organisations and whole structures of jobs or careers. This fifth major kind of use value of science is the focus of a later chapter: Design of Jobs.

The important emphasis to end on is that the uses of science have changed and continue to change as part of a complicated and contradictory process of change in society, within general relationships of power, privilege and work. Changes that we can see over four hundred years since Bacon or three hundred years since Newton, over two hundred years since Wedgwood or a hundred years since Edison and the golden age of classical physics, are all changes in culture at a quite general level. They are changes in authority, ownership, ideology, occupations and status. But they are also changes in leisure and language, the general complexity of lives as they are lived, and the numbers of people and places directly implicated in each and every action.

THE MILITARY AS FATHER CHRISTMAS

MILITARY RESEARCH and development plays a key role in shaping priorities and actual products across the whole range of sciences, technologies and medical practices. This is hardly surprising given that up to half the R&D in 'advanced' countries is funded out of military budgets. Many of us depend on military patronage for jobs, and this is especially visible in the USA. Boeing is the biggest company in Washington State and Lockheed, Rockwell and Douglas between them provide a lot of paychecks in Southern California. McDonnell dominates manufacturing in St Louis. Bath Ironworks is the primary employer in Maine. Thirty-one thousand workers were involved at Groton, Conn., in the Trident submarine, and when Rockwell's contract for the B-1 bomber was cancelled in 1977 it hit thirteen thousand workers directly, plus forty thousand subcontractors' workers.

The lock-in to so-called 'trend innovation' (change without change) in military technology lies partly in the way that prime contractors are tied to products and to arms of the military. Boeing, General Dynamics and Rockwell make bombers, Lockheed makes heavy air transports and submarine based missiles. Chrysler and GM make battle tanks, British Aerospace makes combat aircraft, Fokker makes transports. Some of the subcontractors are themselves very large: Rolls Royce, Pratt and Whitney in aero engines, Texas In-struments and Ferranti in electronics. The small subcontractors are unstable and the mix varies with the technology. In lean years they go bust, and this affects not just their direct employees but also employees in suppliers' firms and service industries. Although direct employment in military production in California is only eight per cent, around forty per cent of the total employment depends on subcontracting, capital goods and consumer goods and services.

With all this dependence you'd think it only right that there should be 'spin off' into non-military products. The US space agency NASA recognises that the space research business has become a kind of Father Christmas by publishing every year a volume of spin-offs. Space research itself was largely inspired by military insecurity following the Russian Sputnik successes in 1957. But military fostering of widespread technical developments is longer established and sometimes less visible. Nuclear physics was a very esoteric discipline in the 1930s. But within ten years it became the largest branch of the weapons industry, and in another ten years the effects had worked their way through in a wholesale restructuring of the heavy electrical goods industry.

The funding of oceanography would not have grown if it were not for submarines' role in the balance of terror. The real growth of seismology is a result of the monitoring of test-ban agreements in the 1970s and 80s. The move from radio valve to transistor to microprocessor was determined by the need for reliable machinery which wouldn't malfunction under the jarring it receives in military airplanes, and then by the demand for smaller and lighter machinery for the same use. The list of similar connections would be very long: computers and World War II gunnery, instamatic cameras and high resolution photography for spy planes and satellites, containerisation and military transportation of supplies.

In the R&D based firms there are people whose job it is to make the connections between military work (the bread and butter) and civil and domestic spin offs (the icing on the cake). But is that how we want our priorities for domestic developments to be set? The issue is not simple. It was from the treatment of air force pilots' burns that immunological investigation led to kidney transplants. The same relationship holds between war wounds and some antibiotics. The largest facilities for basic research in nuclear physics depend on military patronage, and some of the most dramatic and evocative pictures of the earth were transmitted from space probes.

The prime question is not whether we want some or any of these results, but whether we are prepared to continue to accept them as 'spin off'. We're old enough to stop believing in Father Christmas.

6   Tools, Money and Careers

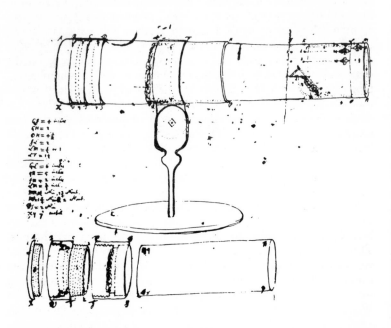

WE MAY OFTEN REGRET the overtly oppressive and repressive 'spin-off' of scientific progress as exemplified by atomic weapons, surveillance electronics or behaviour control through neurosurgery. Although we could use the rhetoric of 'abuse' to name the developments and tendencies which we oppose, it would be a false distinction, based on a false model of the 'purity' of science. Developments such as these are inextricable aspects of scientific work and scientific 'progress'.

The uses of science are not merely contingent. They are the use-value of the knowledge, that is, they are the purposes which have sustained the general interest of the research community and created a place for it in the apparatus of society. Although the connection between research and technique is far more effective today than, say, at the beginning of the nineteenth century, there has always been a connection at a general level between the interests of researchers and the practical needs of commerce and warfare.

HEADPIECE Isaac Newton's own drawing of his reflecting telescope, 1671. (Correspondence of Isaac Newton, University Library, Cambridge)

Newton was interested in properties of metals not just as an experimentalist but also as a functionary responsible for overseeing the minting of coinage, and his interest in optics was part of a concern in England and Holland arising from the practical use of telescopes in surveying, navigation and trade. Less obviously, there was a political component to Newton's mathematics insofar as a major but little-publicised part of his studies was devoted to arguments concerning the necessity of God and the truth of biblical texts, which underpinned temporal authority in seventeenth-century England.

Such listings could go on indefinitely: Darwin's voyage on the *Beagle* as an aspect of imperialist exploration and commercial exploitation in South America, where the British built a large part of the railway system; geological surveying as a process of direct commodity production, since maps of, say, carboniferous deposits had a straightforward market value; cataloguing and analysis of materials from the colonies. These were not incidental interests, but determined the timing and direction of whole fields of natural scientific research, including those which later (and temporarily) came to be called 'pure'.

By the start of the twentieth century, thanks to the extent of socialisation in scientific detail labour, it was possible for financiers literally to buy and sell technical problems and invented solutions. This is what made Thomas Alva Edison a success, where a hundred years earlier Josiah Wedgwood, in a sense, failed. Both had keen commercial sense. Both experimented intensively, employing workers to assist them in looking for saleable inventions. In 1775 Wedgwood had proposed a hundred-member 'experimental work', a collective research and development project for the pottery industry to be financed as a joint-stock company. His fellow businessmen did not want to buy; they were in the business of manufacturing, not researching. By the end of the 1800s, however, technology in the form of techniques and machinery deriving from research was established as a mode of commerce. This made the difference for Edison.

When Edison tossed one of his customary sketchy notes on to the bench of a workman called Kruesi – 'Kruesi, make this. Edison.' – he could be confident that 'this' (as it turned out, the phonograph) would be bought. Technical answers to social questions had become a norm. To be realised, the 'experimental work' of Wedgwood had to wait for the day of imperialism, of Bell Labs, IBM and Du Pont, of IG Farben (the 'IG' – Interessengemeinschaft – a German dye manufacturer's cartel) and ICI (the British response). But when it happened, the experimental work was not on the model of a collective venture of subscribers and experimenters, an expanded and capitalised version of eighteenth-century drawing-room experimentation and popular lectures. It was the perfunctory 'Make this' of the company executive which galvanised experimenters, as paid research and development workers within twentieth-century joint-stock empires in which ownership, management and labour were quite distinct roles.

By 1912 Edison was complaining: 'Long delays and enormous costs incident to the procedure of the courts have been seized upon by capitalists to enable them to acquire inventions for nominal sums that are entirely inadequate to encourage really valuable inventions. The inventor is now a dependent, a hired person to the corporation.' Edison himself was hired by Jay Gould, the financier, to invent a novel electrochemical relay which would enable Gould to attack the Western Union Company which held all patents on electromagnetic relays. R&D (research and development) is part of the normal – capitalist – procedure of business, and it is big business.

Edison's barn laboratory in Menlo Park, New Jersey, was a prototype of the industrial research establishment. Industries built around this institution now dominate the powerful productive institutions of Edison's day – the machine shops of Detroit and Cincinatti. There is a town called Menlo Park in California, on the San Andreas fault in Silicon Valley, where the median income is $30,000 a year. In 1979 companies in the 'valley' did $6.2 billion worth of business.

LIGHT AND FRUIT

The useful distinction between experiments of 'light' and experiments of 'fruit' is Francis Bacon's. It is, however, an analytical rather than a practical distinction, as his successors knew. The seventeenth-century founders of the Royal Society of London recognised that some investigations of intellectual interest (experiments of light) had at the time no perceptible implications for day-to-day commerce or technique. But they were committed to building a knowledge 'for the use of cities, and not for the retirements of schools'. The Society's chronicler, Thomas Sprat, pointed out that although experiments did not always 'handle the very same subjects that are acted on the stage of the world; yet they are such as have a very great resemblance to them' The activities of the

106

Society, which was to be the 'general bank and free port of the world', included 'banking' knowledge – curiosity – and also the accumulation of wealth through the 'spending' of knowledge. Problems of trade and manufacture were studied by those directly engaged in them, and by others who derived incomes from them. Traders meditated on scientific aspects of their problems and 'men of leisure', enthusiasts, pursued inventions which they thought might draw big dividends.

Such intimacy of interests no longer characterises today's research at the personal level, simply because most scientific workers today do not determine what it is that they work on. They work as employees, or as non-autonomous agents of 'progress' in a field. 'Pure' science evolved during the late nineteenth century and disappeared during the mid-twentieth. Throughout, the detail labour has taken place under the general determination of practical and political interests, as perceived by managerial and entrepreneurial elites of senior scientists, industrialists, financiers and state administrators.

Spin off

This sort of connection is quite clear, for instance, in the 'spin-off' from vast, capital-intensive projects such as those within the US space programme. Non-stick surfaces for cooking pans and sticky-label backing papers, welding techniques that use shaped explosive charges, a whole gamut of microelectronic devices from calculators to phone bugs, super glues, lightweight honeycomb-construction building panels; these all flowed from development work financed under the imperative of space research. At one level the imperatives were political, being connected with the US-Soviet arms race and the balance of terror. And at another level the politics were those of jockeying for benefits within a capitalist economy.

At this level 'spin-off' is far too casual a description. It may have been impossible to forsee in detail any of the specific products and processes which finally emerged, but industrial corporations certainly recognised the possibilities of 'fruit' which could grow from

massive state underwriting and financing of investments which individual companies were not prepared to risk. The only real argument for the vast Anglo-French state expenditure on Concorde was that its technological sophistication might prime the pumps of applied research in a more general way.

'Pure' Research

What about 'experiments of light'? The thing to note is that 'pure' research is an archaic label. Fashionable and politically useful during the late nineteenth century, this description served to imply a disengagement at the institutional level between scientific workers who were dependent on public support and sought to increase it, and specific, controversial political or religious interests. This was the age of the great confrontation between religion and science, centred on the theory of evolution. The 'pure' identity has now lost its fund-raising value, and it is usefulness not detachment which counts with funding agencies and patrons. Speculative and general research is now called 'basic' or 'fundamental', the clear implication being that it is basic or fundamental to some purpose, undefined but anticipated. The Baconian principle is more overtly recognised in present-day terminology but it has never really ceased to be active in a practical sense.

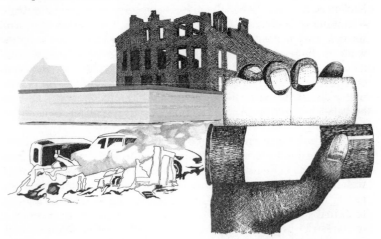

CS gas, Toxteth (England), 1981; plastic bullets, Falls Road (Ulster), 1970s; the relief of man's estate, Francis Bacon (Lord Chancellor to James I of England and Scotland), 1560 . . .

It was 'pure' research which produced the knowledge that was exploited in military form as 'CS' gas, the riot-control agent. Two research chemists, Corson and Stoughton, noted in 1920 that *o*-chlorbenzylidene-malonitrile had 'lachrymatory properties', a fact which sat unused in the literature until development chemists at the

British government centre at Porton in the 1950s turned it into the more prosaic and concrete fact of CS gas. It would of course be unreasonable to argue that Corson and Stoughton were culpable, though they were immortalised in the use of their initials to name the offensive product. They were just engaged in 'normal science', the practice whereby the average academic research worker turns out the average three papers a year documenting the properties and preparations of odd chemicals, thus filling out the map of knowledge with tiny notations. Jigsaw-puzzle science, a genteel pastime, cataloguing the stockroom of nature.

Innocent enough, but problematic at a general cultural level. Contrast this case with Edison who ransacked nature, trying one material after another, in search of a substance which would serve as a durable filament for electric light bulbs. The purpose was explicit and immediate, and the search intense and ruthless, with no false consciousness of 'purity'. Two academics in the 1920s, on the other hand, were merely exploiting the 'professional' mode of work, a meal ticket, a perpetual but by convention purposeless game played against nature. They can hardly be held responsible for CS gas, but as participants in a more leisured version of Edison's ransacking of nature, 'detached' researchers can be criticised. First, they are self-deceiving, in a way that the founders of the Royal Society never were, about the relationship between 'light' and 'fruit' – banking and spending – as use values of scientific knowledge. And second, they perpetuate, as actors, the mystification that scientific work of any kind exists only for the delight and material comfort of its participants. Science has, and always has had, cultural and economic purposes beyond those of its most narrow-sighted and complacent detail labourers, just as other corporate institutions have purposes beyond those of individual employees.

The Manhattan Project
The age of innocence, as far as the role of management in science is concerned, ended with the Manhattan Project of World War II, which produced Little Boy and Fat Man, the Hiroshima and Nagasaki nuclear bombs. Very large numbers of workers of high intellectual standing were employed, together with massive resources of industrial production like those needed to refine uranium ores. Scientists, directors and managers of industrial corporations, and military and civilian state officials were all forced to work closely together by the urgency and secrecy of the project but also by the intrinsic complexity of the apparatus which spanned fields of work that were normally separate.

At a general level the Manhattan Project represents the discovery of the 'mission-oriented' mode of scientific and technical work. Scientific opinion initially generated the awareness that a nuclear weapon was possible, and Albert Einstein wrote to President Roosevelt in August 1939, pointing out that 'an extremely powerful bomb of a new type' might be built. Once this idea had been transformed into a military and political need, questions of technical possibility and natural knowledge became entirely subordinated to the goal. Objections of scientific and professional principle were squashed by weight of numbers and capitalisation, the sheer unwieldiness of a massive political and economic apparatus. 'Objections' of nature, the disorder of Nature seen as a storehouse of resources, were dismissed by the concerted efforts of an army of detail labourers, forcing open each black box of natural order by throwing at it the sheer weight of a massive economic and cultural apparatus.

Great Men of modern physics were caught off balance when less brilliant colleagues like Robert Oppenheimer and Edward Teller began to rise high in the effective hierarchy of the project. They rose as managers. Scientists were not in fact central in the project. They were crucial, as technical personnel who could make the Bomb go bang, but peripheral in relation to the apparatus of decision-making. Morally critical scientists may have impugned Teller's intellectual status; but this was a side issue, for Teller had power. This kind of intimate and contradictory relationship between those who do the work and those who control (contradictory, because Teller was a worker too) is a main focus of any history of science as production. Our ordinary experience of work is of just the same kind. Great Men of science are historically and, in the present, institutionally remote from our direct experience. But the Grey Men of management are only too familiar, which is why this history of science has to be a history of science as work.

Big Science and Spending

Apart from its role in establishing nuclear weapons and nuclear energy as new major areas of state-controlled research and development, the Manhattan Project set a new model for 'pure' science which dominated many sectors of post-war research. To some extent this was because the personnel of the Project were themselves dominant. A fourteen-man Science Council proposed by US President Reagan's administration in 1982 included Edward Teller and two other Manhattan Project scientists among its several physicist members and contained no women, one biologist and no social scientists.

Bigness became the mode of post-war basic research in high-energy particle physics ('atom smashing'). The path has been: more powerful forces, and therefore bigger particle accelerators, and therefore more expensive plant, and therefore more management. Today's high-energy research uses plant which consumes electric power on the scale of a small town and is staffed like an industrial site. National and international coordination of experimental schedules is necessary (many British researchers use facilities at CERN, the European Centre for Nuclear Research in Geneva) so that the capital-intensive equipment can be utilised to the maximum. Just as in industrial shop-floor work, a three-shift system operates for experimenters. Large research groups are formed which have within them a tight and minute division of labour. A young post-doctoral researcher may in fact be just an electronics expert, another may specialise as the team's computer programmer, a third in managing the women (perhaps a hundred in a 'team' of a hundred and sixty) who scan the hundreds of thousands of scratchy-looking photographs of particle tracks which are the experimental products. This is the apparatus of the biggest Big Science.

It is not just 'pure' science writ large, because the massiveness of the apparatus enforces a high degree of planning. Even basic research is turned, through budgeting, construction costs and experimental scheduling, into mission-oriented form. Professors in this field of physics have been known to sell it on the strength of the training that it provides for people who may later wish to move into industrial management.

Construction costs for the largest high-energy accelerators run into hundreds of millions of dollars. The US National Accelerator Laboratory cost $250 million, the Stanford Linear Accelerator Centre (in Silicon Valley) $113 million, the Brookhaven Alternating Gradient Synchroton $30 million, which is about the same as the original CERN proton synchroton. These establishments also cost millions of dollars each year to run: NAL – $36 million in 1975, AGS and SLAC – $25 million. Advances in research, however small, depend on the power and therefore on the cost of accelerators. Owen Chamberlain and Emilio Segre won a Nobel Prize for discovering the antiproton using such a machine, the Bevatron. Cynics suggested that it was not so much the experimenters' skills which led to the discovery as the fact that the machine had for the first time summoned up enough energy to produce an antiproton. Should the Nobel Prize have gone to the Bevatron? Or even to the taxpayers whose money supported it?

Less spectacular concentrations of capital equipment do not necessarily mean that work in other areas of basic research is less determined by money. Telescopes are expensive, and since optical telescopes need to be built in remote places where the atmosphere is clear, international coordination and funding is intrinsic to the science of astronomy. Oceanography also calls on sophisticated equipment and international coordination. Even within the chemistry or biology laboratory, where research may take place on a smaller scale, capital-intensive techniques are used. Chromatography used to be an area of research itself, involving skilled manipulation of temperamental equipment but now, thanks to that earlier accumulation of know-how, fully automatic equipment performs chromatographic analyses on a routine and continuous basis, serviced by unskilled maintenance workers. The more expensive items of equipment bring with them shift work; and this applies to computers which are now an indispensable research tool in the labs where the researchers work.

Light or fruit, science is a force of production. Whether by its perceptible contributions to the economy in industrial R&D, or by the extent and complexity of its own internal time-economy, modern science is overtly an industrial form. By the way that finance determines the direction and organisation of work, it is plainly a capitalist industrial form.

SCIENCE AS A FORCE OF PRODUCTION
The connection between experiments of light and fruit, difficult to map and hard to document statistically, has become part of the ideological middle ground of contemporary politics. By the 1960s science accounted for around two per cent of Gross National Product in industrialised countries, which made it too big to be left alone. In Western capitalist countries, the Organisation for Economic Co-operation and Development (OECD) was initiating a kind of Marshall Plan for science. In Eastern Europe and the USSR 'the scientific and technological revolution' was becoming an ideological fact, eclipsing socialist revolution.

In Britain the 1964 Labour government proclaimed the 'white heat of technological revolution' in the spirit of the new orthodoxy, with a general commitment to establish direct ministerial planning responsibility over civil science, to strengthen connections between wilfully or unwittingly detached academic research and industrial practice, and to oversee industrial rationalisation. There was a long-standing tradition of socialist-Baconian rhetoric which spoke of science for the relief of man's estate through the production of more and better goods. In the early post-war years, 'consumer research' was one focus for advancing the idea among Left scientists. The ideal which became firmly entrenched in practice was that of science for the production of goods within a 'mixed' or capitalist economy. This meant science for profit and for the reproduction of military, industrial and financial establishments, suitably 'rationalised' through state aid, state corporations and state-prompted mergers in electrical goods, cars and trucks, aerospace industries and computers: General Electric (GEC), British Leyland, British Aerospace and International Computers (ICL).

Responses of scientists to the changing circumstances of their work have been prolonged and confused. By 1968 organisations like OECD had found it necessary to draw up formal definitions embodying distinctions foreign to the traditional consciousness of scientists: pure basic research, oriented basic research, applied research, experimental development. Science in the 1960s was manifestly a more economically central and managerially ordered activity than

U.K. STATE FUNDING of R&D 1980

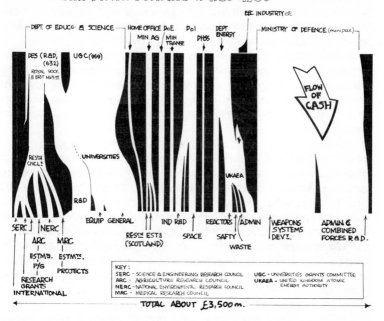

EE INDUSTRY etc.

DEPT. OF EDUCN. & SCIENCE — HOME OFFICE DoE · Pol · DEPT ENERGY — MINISTRY OF DEFENCE (minipax)

MIN. AG / MIN TRANSP · PHSS

DES (R&D) (632)
ROYAL SOCY. & BRIT MUSm

UGC (969)

FLOW OF CASH

RESH CNCLs

UNIVERSITIES

UKAEA

R&D

SERC · NERC · EQUIP GENERAL · IND R&D · REACTORS · ADMIN · WEAPONS SYSTEMS DEVt. · ADMIN & COMBINED FORCES R&D.

ARC · MRC · RESm ESTs (SCOTLAND) · SPACE · SAFTY · WASTE

ESTMts. · ESTMts.

P/G · PROJECTS

RESEARCH GRANTS INTERNATIONAL

KEY:
SERC – SCIENCE & ENGINEERING RESEARCH COUNCIL
ARC – AGRICULTURE RESEARCH COUNCIL
NERC – NATIONAL ENVIRONMENTAL RESEARCH COUNCIL
MRC – MEDICAL RESEARCH COUNCIL
UGC – UNIVERSITIES GRANTS COMMITTEE
UKAEA – UNITED KINGDOM ATOMIC ENERGY AUTHORITY

← TOTAL ABOUT £3,500m. →

US STATE FUNDING of RESEARCH 1980

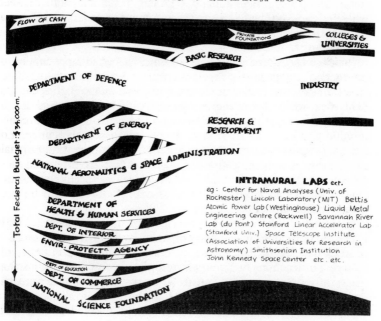

FLOW OF CASH

PRIVATE FOUNDATIONS

COLLEGES & UNIVERSITIES

BASIC RESEARCH

DEPARTMENT OF DEFENCE

INDUSTRY

DEPARTMENT OF ENERGY

RESEARCH & DEVELOPMENT

NATIONAL AERONAUTICS & SPACE ADMINISTRATION

DEPARTMENT OF HEALTH & HUMAN SERVICES

DEPT. OF INTERIOR

ENVIR. PROTECTn. AGENCY

DEPT. OF EDUCATION

DEPT. OF COMMERCE

NATIONAL SCIENCE FOUNDATION

Total Federal Budget: $34,000 m.

INTRAMURAL LABS ect.
eg: Center for Naval Analyses (Univ. of Rochester) Lincoln Laboratory (MIT) Bettis Atomic Power Lab (Westinghouse) Liquid Metal Engineering Centre (Rockwell) Savannah River Lab (du Pont) Stanford Linear Accelerator Lab (Stanford Univ.) Space Telescope Institute (Association of Universities for Research in Astronomy) Smithsonian Institution John Kennedy Space Center etc. etc.

was recognised in conventional images. But the conventions were powerful. When Lord Rothschild proposed, in *A Framework for Government Research and Development* (a Government committee's report of 1971) that customer-contractor research should be the norm in financing scientific research, there was an outcry in defence of 'academic freedom' and creativity in research. Researchers today, sensitised by government spending cuts, queue for this type of contract funding.

A Body of Knowledge?

In thinking about modern science it is important to understand what is meant by a force of production. It would be a mistake to think of science as a body of knowledge, the 'light' produced by a somehow detached research community, which is then available for application to yield 'fruit' in the form of technology and technique. For one thing, basic research is largely integrated into industry. In the USA more basic research is funded in industry than in colleges and universities. If it is a 'body' of any kind, science is a body of practice, an organised collectivity of people, materials, machines and purposes, which is connected at many levels with other practices in society.

Science is thus only analytically separate from, say, manufacturing, politics, consumption or education. As a form of practice, science in the eighteenth century was marginal – economically, numerically and intellectually. Science now is central, as a sector of public and private spending and as an intellectual domain. It is this whole relationship between science, economy and culture which we should have in mind when we talk of science as a force of production.

People who use the phrase often reduce the meaning of 'forces of production' to machines. This is a crude interpretation, which only seems more crude when it emerges in discussions of science, for here it is obvious that the intellectual content of the practice is central, as are the practical connections between a research 'community' and other social groups. Money and tools, though crucial are only elements of the total relationship. The complexity of the relationship is no justification for managers and state functionaries who, when dealing with science policy issues, try to solve the problem by throwing money at it. As a practical consideration, quality of relationships is prior to quantity of things.

Nevertheless, the complexity of science as an assemblage of things is considerable, and is the basis of some of the distinctive characteristics of present-day practice. We can talk about this aspect of science by speaking of its *apparatus*. This refers not just to the pieces of equipment used, but to all of the physically identifiable things which are connected together in practice: equipment, buildings and energy sources, workers themselves, journals and, less

114

tangibly but measurable and absolutely central, money.

When modern science is referred to as a force of production, the emphasis is on three things:

□the geographical extent of the apparatus (science is an international – or multinationals' – practice)

□the density or intensity of the apparatus (the centralisation and concentration of research activity) and

□the manifold connections between the apparatus of science and other cultural, political and economic institutions (medicine, manufacturing, schooling, government).

The extent and complexity of the apparatus of modern science, and its entrenchment via these cultural and economic connections, is what determines the importance of management in science generally, and 'mission-oriented' research in particular.

MARKET VALUE: CHIPS AND GENES
Science has entered the market. The market has entered science. We can like it or lament it or maybe change it, but as an issue we cannot avoid it. As this chapter (indeed, the whole of the book) stresses, science has always been determined in its content and direction and style by values. This section sets out some of the forms of market value in science using illustrations mainly from chips and genetic engineering (robots and satellites and bugs), two of the most potent areas of current research and development and investment. Market values are linked to the form that things take as commodities. Four distinct ways in which commodity forms figure in the work of scientific workers are profit, patents, wage work and career-capital.

Profit: Producing Commodities
In the robots and satellites sector the British government financed the setting up of three companies (INMOS in chips, INSAC in software and NEXOS in office equipment) to compete in this sector of the international market. The pace of innovation in this area is unusually fast, and the high capital investment required to enter the sector as a manufacturer together with the possibilities of returning a low rate of profit place a premium on agility. But special characteristics aside, this is just one more in what is now quite a long line of R&D based industries. Chemicals is another. Although we have lived with it for many generations now, the chemicals industry really came into its own with World War II and new products such as plastics. The major potential in the coming decades is less in products than in processes based in biotechnology. The British government's policy on research in this sector is straightforward: let private capital do the funding. (INSAC and NEXOS may also be sold off to private buyers.)

Much research in gene splicing is done in academic laboratories.

Yet on looking closer it is possible to see a radical development in the relations between academic and profit-making research, for in this rapidly growing market many of the firms have been started by scientists leaving academia – a hidden State subsidy to private capital. One such firm is Genentech which markets insulin, an expensive but necessary drug (diabetics rely on it) which can be profitably made through genetically engineered bacteria. Another such firm in the Cetus Corporation, which has gone public, and its shares have made millionaires of its owners. In these cases scientists went to the market. In other cases firms already established in perhaps the pharmaceuticals business moved in on gene splicing research: Imperial Chemical Industries, Glaxo, Unilever and many more of the chemicals multinationals are now in the competition.

Chips and bugs have become the basis of commercial bonanzas while their academic forebears, solid state physics and molecular biology, remain relatively academic. Circumstances differ. The history of solid state physics and microelectronics was commercial from an early stage. The world's largest private industrial research institution, the Bell Labs of AT&T (American Telephone and Telegraph), with a little help from its military friends, generated the work from which transistors, then chips and microprocessors developed. William Shockley, who set up his shingle in Santa Clara Valley, California, in 1956 and became 'father of the transistor' was an ex-employee of Bell Labs. Molecular biology, on the other hand, was almost entirely funded by public money through the National Institutes of Health in the United States and the Medical Research Council in Britain. Decades of publicly funded research have now been appropriated by private capital.

One particular consequence of aggressive market competition is that it generates pressure to soften government policy on safety controls in rDNA research. The head of the Cambridge Laboratory of Molecular Biology, Sidney Brenner, commented in 1977, 'If there are to be economic benefits coming out of this work, it would be a tragedy if the United Kingdom was not there to reap some of these.' In particular, 'It would be a great pity if in the end this country found itself paying licensing fees and royalties to other countries for end products.' This is not so obvious as Brenner makes it seem, since shrewd multinationals like ICI have in recent years chosen to opt out of some lines of manufacture and research and buy in the expertise or products, in order to concentrate on areas where they feel most certain of profits. The general implication is plain, however: British research and manufacture must not be impeded by cautious if well-intentioned State agencies.

This is only a fantasy, but consider the possibility that some of the world's most authoritarian and corrupt regimes in Third World countries might offer, not a tax haven, but a *hazards* haven for firms doing rDNA research. The firms would get control-free experimentation, while the heads of the regimes would get a more stable economy. After all, Puerto Rico played the role of providing women for early research on the contraceptive pill when risks were unclear.

Patents: Knowledge as a Commodity

Knowledges are determined by property rights and a market place, through the patent system. Hence knowledge itself is a second location of commodity relations in science, as distinct from any products (such as insulin) which may be made using that knowledge. Some companies like General Chemical and General Electric were created expressly to control patents. A lot of effort goes into research to enable firms to get round their competitors' patents or to make their own patents watertight. A disaster at Flixborough, England, in the mid-1970s, when a chemical factory making an intermediary in nylon manufacture exploded, was able to happen because the process route, which had intrinsic technical weaknesses, had been selected by the company to escape from a spider's web of patents. Research in the microelectronics industry is centred on patentable discoveries, but the high security maintained in Silicon Valley is not entirely due to industrial espionage – the KGB is rumoured to operate there too, on behalf of the backward Russian computer industry. Some of the research areas of Cetus Corporation are high security, not because of microbiological hazards, but because of potential industrial espionage. Techniques and materials represent such a high investment of research time and such a huge potential market value that the Corporation will not even patent them because this would give too much away.

The patent on the basic system used in genetic engineering which is based on a modified variety of *E. coli*, a common bacterium, may possibly be the most lucrative ever taken out. Two applications for patents on such techniques were turned down by the US patent office for a variety of reasons which included 'You cannot patent a product of nature', and hinged on the technical point that living matter was not explicitly included in the wording of patent law. In both cases a patent appeal court overruled the decisions and in both cases an ordinary appeals court upheld them. The manufacturer dropped one case because it might obscure the issue and took the other to the Supreme Court where it won, but only by a very narrow (5–4) margin. Not only is the patenting of life a legally questionable practice, but also the private ownership of life forms means that they can be denied to those who need them.

Plant Breeding Rights illustrate this powerfully. PBRs are the plant breeder's patent, and entered British law in 1965. During the following week Rank Hovis McDougall (which owns flour mills, bakeries, food manufacturers such as Atora and Paxo and Saxa Salt, wholesalers, transport, finance companies and animal feed suppliers through sixty-four overseas subsidiaries) bought eighty-four county seed supply companies. In the US eighty per cent of PBRs are owned by four companies. First World plant breeding concentrates on high yield (per unit of capital), which is not the need of most Third World farmers. But they must buy their seed, increasingly, from the First World multinationals – and high yield varieties need fertiliser, herbicide, fungicide, etc.

Wage Work: Labour as a Commodity

The third area of commodity relations in science is that which it has in common with the majority of workers' work – the selling of workers' skills, know-how and time. The consequences of this particular historical way of socialising scientific work are widely discussed in most of the chapters of this book. Scientific workers sell their know-how in two broad ways, first under wage-work contracts as employees of an employer, and second as 'private enterprise' services. The widespread and increasing practice of short-term contracts for funding research directed at specific named results comes under the first heading. Under the second comes the long-standing role of the private medical practitioner, as well as other types of consultatory practice, e.g. in engineering and in the design of computing systems and software. A consultancy, Biogen, markets the collective expertise of leading academics in the field of biotechnology research, who presumably also act as brokers for the labour of the workers on their staffs.

Pressures on workers through short-term contracts can be particularly severe. The funds tap is easily turned off after only a year or two of work, and the whole funding policy of granting agencies is designed to prevent researchers from continuing on short-term grants into middle age, forcing them into a decision to take a tenured post in their early thirties. A research worker can thus be confronted with a painful choice: either to follow the funds and leave the area of work which they value, or to stay with the research and risk being out of a job. This is an increasingly frequent dilemma.

Career 'Capital'

In the fourth category past achievement is exchanged against present and future status. Achievement accumulated in this way is referred to as 'career capital'. Here the commodity relations do not require an explicit exchange, of money for a commodity such as a product, a package of information or a commitment of effort. This is a metaphor rather than a literal meaning of 'capital', although the distance between the two can be rather small. Grants, contracts and academic tenure are financial rewards of different kinds which are dependent upon some measure of accumulated achievement, but in a rather abstract way. It often does not matter exactly what the work has been so long as it was published in a reputable academic journal. The continual pressure to publish prevents many academics from doing what they would consider real work which they personally value.

Grants, contracts and tenure are part of the patronage system in science. Private patronage, such as that which Newton or Descartes or Hooke relied upon, is no longer the norm in scientific work. But the 'fixer', the grant-swinger, is a key figure in the scientific 'community', a broker of money and influence. Military funding is particularly important in this connection. Quite a large part of military-funded research has no apparent value for military purposes ('light' rather than 'fruit'), but what matters to paymasters almost as much as useful results do, is the loyalty and influence which can be bought through the career hierarchy that constitutes a large, prestigious, heavily funded project. Funding helps to 'convince' prospective research students to work on particular problems, and individual scientists or research students often naively attribute the funding of their work to the research director's scientific prestige, rather than to their role as a procurer of commitment. Like industrial or State support for research generally, military funding is partly an investment in workers' intangible but politically real sense of obligation to 'the firm'.

Obligations are created and strengthened through other career means too, such as writing references. When a research director or professor's letter of reference will make the difference between getting or not getting tenure, a job or a grant, not many scientific research workers are in a position to resist 'suggestions' for lines of research. It is notable that famous researchers often have famous pupils; Berthollet, Gay-Lussac, Magnus, Clausius and Helmholtz, Gibbs, Planck and Bernal, Perutz; or among Nobel Prize winners Fermi and Chamberlain and Segre, or Nernst and Langmuir and Millikan, or Schrödinger and Pauling. While this must have something to do with apprenticeship it must also have something to do with patronage, especially when so much of modern research (such as that of Chamberlain and Segre on the antiproton) hangs on access to expensive equipment in great demand. Research workers are free to choose their line of research within definite limits. Progress in science is very closely tied up with progress through career hierarchies, from positions where money is a way to get to do research, to positions where research projects are a way to get money.

Competition to get grants and contracts is part of life in the 'community' of science. A woman worker in cancer research points out, in June Goodfield's account of laboratory life, *An Imagined World*, that 'Everywhere in science the talk is of winners, patents, pressures, money, no money, the rat race, the lot . . .' A good illustration is James Watson's first person account of the race to determine the structure of DNA, *The Double Helix*. The cutting edge of competition is sharpened, however, by the emergence of real capital (as distinct from career 'capital') in a field such as

119

biotechnology research. Taking to the commercial vocabulary quite unselfconsciously, researchers protest against what they see as over-stringent safety legislation by saying 'You'll put me right out of business', when what they mean is that their less-fettered risk-taking competitors may get to the patentable discoveries first. The fact that there is a commercial as well as a career market in bio-technology knowledge and know-how means that research workers may be able to choose to emigrate in order to stay at the research front, in countries whose commercial interests are able to secure a more relaxed state of legislation.

Capital then becomes identical in practice with career capital, while safety becomes a drag on both markets.

The market in DNA knowledge is so new and lively that even the traditions of fair play in the City have not got their analogues in research. The City discriminates against 'insider trading'. Yet it is possible that the same senior research scientist can simultaneously be: i) a member of a committee of the Royal Society which recommends slacker guidelines for the regulation of rDNA research; ii) a major witness submitting evidence to that committee suggesting that serious risks are unlikely; iii) head of a laboratory likely to benefit from relaxed controls; iv) working on a research front where lucrative patents are expected; and v) a member of a government watchdog committee advising on public policy in genetic manipulation.

With the market so visible in an area of science it was surprising to hear mandarin scientist Lord Todd, Nobel Prize-winner and President of the Royal Society, pronounce in 1980 that 'I am wholly opposed to any attempt to regulate or control the direction of scien-tific enquiry and I believe that in saying so I speak for the Royal Society.' What else were the University Grants Committee, the Science and Engineering Research Council and the Medical Re-search Council doing, when they vetted applications for funds in 'peer review'? What else were corporations doing in the field of Lord Todd and some of his mandarin peers, rDNA research? What difference could he see between the current convergence of privi-lege, patronage and private capital, and the possible directions pre-ferred by the Genetic Manipulation Advisory Group (GMAG), the target of his attack? Perhaps some members of GMAG couldn't be trusted to play the game. In addition to the usual members of the scientific elite who sit on advisory bodies, GMAG had trade union representatives with no career interests to protect and no patronage to purchase favours with, only their jobs and their safety at work and their interests as citizens. As it is run, these are not the values which govern science.

Managers Must Measure

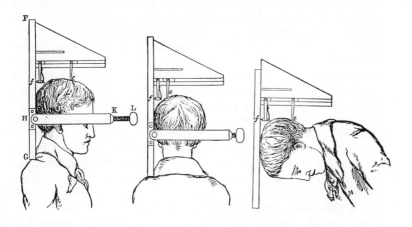

IN THE LAB AND IN THE FACTORY AND IN GOVERNMENT experts measure. It is preferable if people can be taken out of the loop, so that instruments do the measuring – that is, produce numerically coded outputs, which people then interpret. But however it is done, 'man must measure'. In science measurements are the raw material of knowledge, and without measurements a science is fatally 'soft'. In the production manager's office, the board room and the corridors of power from Whitehall to Washington and the world over, statistics are the persistent reference point. Statistical techniques came early in capitalist society as mortality and morality came under official scrutiny for purposes of profit and political administration: insurance, annuities, raising revenues for the State, controlling population and endemic disease. Actual voices serve to mildly qualify the statistical picture, or give it local colour; but even in courts of law, eye-witness evidence now has secondary status in relation to the measured evidence of forensic scientists.

At the centre of the changes in thought of the sixteenth and seventeenth centuries, which we call the Scientific Revolution, was a commitment to seeing the universe in mathematical terms. Galileo asserted that the book of nature was written in the language of mathematics. Newton based his theoretical work on the belief that the project of natural philosophy was to explain all there was in heaven and earth through the fundamental measurables, the quantifiable qualities of matter and motion – size, shape, distance and hardness. This metaphysical commitment, to sticking numbers on things and then explaining their behaviour by reducing the numbers to abstract law-forms written in algebra, has become a dominant aesthetic of modern science – an ethic, even – and a dominant process in the modern management of social affairs.

The ideal of the quantifiers' strategy has been very fully achieved over three hundred years – that all experience and social relations should be valued only to the extent that they can be coded and processed within an 'objective' world of mathematical reason. Down this slippery slope have gone, first of all, colour, taste and smell – very early on, these. Terrible, *subjective*, soft qualities. After them went temperature, sensation, perceptions, ideas, dispositions, skills – the last signifying a whole history of scientific physiology and psychology. Still being sucked in are accomplishments, aspirations, wishes, needs, choices, statuses; and so on and on, into the whole repertoire of contemporary statistical techniques of demand- and opinion- and work-measurement and management. The whole expanding project of science lies in the reduction of experience. Though the outcomes are the same so far as everyday practice is concerned, there are ways and ways of doing the reducing, and it is as well to be sensitive to what they are.

REDUCTIONISMS

One kind of reductionism insists that there is only one single set of theoretical terms ultimately available for describing all meaningful behaviour and action. We might call this *ontological* reductionism (ontology is the naming of what reality is fundamentally made up of – for instance, one pre-seventeenth century ontology had the universe constituted from forms, essences, natural places and purposes). It is this kind of reductionist project which Newton and his contemporaries set in train when they tried to analyse the cosmos as mere matter in motion. No 'spirits', no mysterious animating forces, just small hard pieces of stuff and God. The project survives and thrives in modern physics, in which sub-atomic structural elements (electrons, neutrons, photons and other denizens of this insubstantial but measurable world) are understood to be the building blocks of all legitimate explanations of physical behaviour. The project has been taken up over the past forty years, with staggering success, in biology as *molecular* biology, which seeks accounts of biological phenomena such as inheritance of characteristics in descriptions of molecular structure and processes.

This most radical reductionism is limited to explanations within sciences so that – in practice – one science does not get reduced to another. There is, however, a sort of peck order of the sciences, with the most radically reduced explanations at the top and the most-nearly experiential at the bottom. The lower down it is, the more chance there is that a science will be accused of being tainted with social, political and ideological assumptions. Economists will say it of sociologists, psychologists of sociologists, ethologists of psychologists, physiologists of ethologists and psychologists, biologists of *all* the human and behavioural sciences no matter how hard they try to pin numbers on things. But molecular biologists have the laugh on the biologists – and physicists sit pretty in the 'purest', most reduced domain of explanation. Mathematics, it should be added, like anthropology and systems theory, seems to be out of the competition, preferring to refine systems of conventions rather than produce explanations of actual happenings and potentialities.

Hardly anybody tries to explain, say, economic behaviour in terms of molecular interactions – which is not to say that scientistic economists don't try to get some of the kudos of chemistry to rub off on economics by using mathematical 'models' from chemistry to talk about economic behaviour. Thus Charles (later Sir Charles) Goodeve, Director of the British Iron and Steel Research Association, Fellow of the Royal Society and later a Director of the Industrial and Commercial Finance Corporation, speculated in the journal *Nature* in 1948:

> We can look forward to the day when the *science* of economics will be sufficiently developed to enable man to gain better control over his circumstances. I believe this science will only grow by the methods of operational research and more specifically by developing its own versions of the First and Second Laws of Thermodynamics, the First being the law of conservation of equivalence, the Second the law of direction of changes as governed by (economic) forces and the associated conditions of equilibria.

There is a clear and general assumption that the explanations of the more rarefied sciences are 'better' than the others, more measurable, 'harder', more fully mathematicised. The relationship between the sciences is seen in ideal-historical terms, as an edge of objectivity gradually advancing from studies of the heavens, through terrestrial physics, into physiology and geology, the history of life, the functioning of mind and the life of society. The pure light of quantitative reason hasn't swept all the way across yet, but the assumption is that one day it will, revealing the hard truth about nature and society.

Despite the seductiveness of ontological reductionism – it is certainly very elegant in an abstract way – most established sciences do not in practice try to go the whole hog. Chemists do not try to cast all their explanations and descriptions in terms of sub-atomic processes. Experimental psychology, physiology, geology and other

123

sciences with a manifestly physical interest in the world are similarly modest. Even molecular biology stops short at the molecular level and leaves particle physics' terms to particle physicists. Physiologists study 'irritability', 'contractility' and other phenomena which are meaningless in the language of sub-atomic particles. Chemists study solubility or classify substances using terms such as 'hygroscopic', which are not reducible to a set of terms taken from physics: and they use them without embarrassment as the everyday practical language of their work, closer to non-specialist language. Without delving into the superfine structure of matter, respectable sciences have been developed around quantitative 'laws' summarising observable behaviour in all these spheres of research.

Mathematics Needs Idiot-Proof Rules
While they accept the practical autonomy of different levels of description and explanation, however, these sciences are reductionist in a different way. By defining their vocabulary – contractility, solubility, and so on – in terms of the conditions under which these qualities can be turned into quantities (i.e., measured) sciences reduce language to a code rooted in numerical operations. There are, in fact, two closely related reductions taking place here. One is a mathematical reduction, a substitution of an artificial, formal language for a 'speakable' language, gaining power in manipulation and quantitative logic but sacrificing translatability into other forms of language, which people habitually use to relate their experiences and to make fine qualitative distinctions. The strategy, with great deliberateness, cuts off scientific explanations from the normal cultural processes of valuing and locating experience. Because of this disjunction between everyday and formal languages, the mathematical reduction requires, practically, a definition of the conditions in which a given measurable is to be agreed to be measurable. It requires, in other words, an operational reduction – a specification of what operations will be counted as adequately pinning down a quality in quantitative form, so that other experimenters may use the same unreal language to relate to the real world. Operational reductionism is intended to make reality stand still for long enough for objects to be singled out and have numbers stuck on them. Because the mathematical and the operational reductions necessarily go together we might refer to them together, as *methodological* (as distinct from ontological) reductionism.

This is universal – in fact it is taken as the entrance requirement – in science, and this is why, despite the ontological differences between the physical sciences, they can share a practical language. The Royal Society standards concerning quantities, units and symbols (1975) state that: 'each physical quantity shall be given a name and a symbol which is an abbreviation for that name. By international convention seven physical quantities are chosen for use as
124

dimensionally independent base quantities: length (1), mass (m), time (t), electric current (i), thermodynamic temperature (T), amount of substance (n), luminous intensity (Iv). All other physical quantities are regarded as derived from the base quantities.' These, plus the ten digits 0–9 and the units of measurement, are the sum total of the legitimate language of observation in hard physical science, and the stuff of all explanation.

Don't Be Soft About Society

The consequences of methodological reductionism are particularly dramatic and visible in the management sciences, where it is the behaviour of humans in society which is being measured and mathematically modelled. The fetish of quantification means that if studies of social behaviour are to be really scientific, on natural-science lines, then the whole of the vocabulary of power, purposes, values and identity must be operationalised, rammed into measurable forms. Safety becomes measurable quantities of toxic compounds. Efficiency becomes profit ratios or yield factors. Morale becomes frequency of attendance at work. Cooperation becomes two parties converging in the same observable behaviour, and values become the results of a statistical analysis of past behaviours. Society passes into this science – hard-form science – stripped of colour, perfume, taste. Fact and value become separated, but only so that value can be reduced to fact. Mind and body become separated so that mind can be reduced to body.

Half a century ago Karl Mannheim, founder of the sociology of knowledge, wrote in *Ideology and Utopia* about the quantitative trend: 'It is not to be denied that the carrying over of the methods of natural science to the social sciences gradually leads to a situation where one no longer asks what one would like to know and what would be of decisive significance for the next step in social development, but attempts only to deal with those complexes of facts which are measurable according to a certain already existing method. Instead of attempting to discover what is most significant with the highest degree of precision possible under existing circumstances, one tends to be content to attribute importance to what is measurable merely because it happens to be measurable.' This is a description of most quantitative social scientists, most OR scientists, most economic planners. Frightened to be anything other than scientific they hamstring themselves with technique. That might not matter if it were only a game or a private pursuit, but these technicians' hangups have more serious consequences.

PREDICTION AND CONTROL

Hard sciences write statements about the world in algebraic language, mathematical models. They turn the handle of the logic machine, and out come other statements, transformations of the models. These are interpreted as predictions and decoded back into the real world using the operationalist codebook, and what you have now is experimental designs, traps set to see if reality will spring them. It works? Then what you have is control – not 'truth', because the models are partial, the experimental conditions limited, the coding and decoding always botched and fudged to some extent, never as mechanically perfect as would be ideal. Truth is not what matters. Control is all, experimental control, the ability to make reality jump through your hoops, to pre-guess the outcomes of confrontations with nature.

Again, this might not matter too much if it were merely a matter of private games played in the artificial world of laboratories (although the morality, the baiting and trapping and out-gunning of nature, surely wouldn't stay within the walls of the scientific workplace. Presumably it wasn't born there anyway). The models and the methods are taken out into the real world and this has two main consequences. The first is power over nature 'in the wild', limited power but quite real and effective. Antiprotons can be forced to show themselves, genes can be mapped and manipulated, a light brighter than a thousand suns can be generated over a Japanese city, non-natural life forms can be made to live and reproduce. This power passes all ordinary understanding, and that is its price – not a necessary price, but the going rate. The partial power to manipulate nature is the power of partial institutions, elites, in government and business. They pay for the work, they own its results, and they have a double guarantee because the knowledge *cannot* be in general circulation. It is not that kind of knowledge. It is not formulated in natural language, and to go between the artificial and the natural language – the language of theory and the language of practice – requires a specific and long apprenticeship and a specific and thorough enculturation. Scientific knowledge is not – declines to be – a part of common culture, it *bypasses* the social processes of ordinary understanding. It need not. But scientists enjoy the games, the competition for experimental control is too intense for them to take time off to open the codebooks to the public, and the owners of science enjoy the benefits of private exploitation of a cultural resource which has never been properly socialised within the culture at large and never will be as long as they keep the pressure up.

The Mathematical Fix

The second real-world consequence of the cultural form of science stems from the transference of the form into the management of society directly. Operational Research is one place where this happens. In World War II natural scientists recruited themselves into the war effort and began to work on 'operational' problems rather than the technical problems of the small back room. They studied why submarines escaped bombers' depth charges, and whether it

126

would win the war faster if resources were committed to bombing submarines rather than German cities, and which bombing strategies would kill most people and destroy most forces of production. This is how one of the wartime innovators, P. M. S. Blackett, socialist, professor of physics, Nobel Prize winner, later President of the Royal Society, saw the thrust of science in war:

> Under the name 'operational research' scientists were put into a position to study and analyse the planning and operational activities of the military staffs, thus encouraging numerical thinking and avoiding running the war on gusts of emotion.

No gusts of emotion, but an unyielding wind of the reduced emotionality of reductionist science.

OR scientists build models, incorporating quantified variables, of the structure or the processes of the systems that they are permitted and paid to study. They build the models, test them for internal formal consistency, try out their predictiveness on past data and then run them under different conditions to see what possible variations in outcome there are. In other words, they use models as surrogate realities, to experiment on. As a result of these thought-experiments they – and of course the managers they service – decide what 'experiment' to try in the real world, that is, what changes to make in practice to achieve a desired outcome: saving in stock-holding costs, reduction in manning levels, 'rationalisation' of production across a number of sites. As they go beyond technique into reality they are making concrete the power that is implicit in pre-conceptualisation, as a general relation of production which separates knowledges from working people. OR scientists and managers think through, in detail even if only partial detail, a course of action before the event, so that the system they subsequently act in and transform has no involvement in the prethinking process. This is what abstract knowledges are for. 'The system', the whole complex of practices – people, machines, money; apparatus, purposes – is only an experimental object, cast in that role by the investigators. Only its ghost, the mathematical model, is present in the preconceptualising process. That need not be so (why not?) but it is.

Thus, following the natural-science pattern of prediction and control, scientists and managers stand apart from the system of which they *are* a part (or they could not influence it at all). They reach conclusions about outcomes of action in real practices without engaging in the full extent of its real-time working and submitting themselves to its cultural conditions. Modellers work in privileged conditions – of access to information, of a flexible time economy which allows thinking (they are *paid* to think), of access to the means of putting thinking into practice. In this way preconceptualisation, rather than being the general human capacity to think first and act afterwards, takes the historical form of *fiat*, the managerial power to act first and (depending on how partial the model's mapping of reality is) successfully. Management science is not simply sticking numbers on things, as its practitioners and pioneers naively put it. It is a process of sciencing society, freezing history to a standstill as data so that 'things' can be picked out and have numbers stuck on them, then moving through models into the practice of elite control, to make the things jump through designated hoops. This is the mathematical fix. It is the power of preconceptualisation. It is a sexist truism, and a mystifying overgeneralisation, to say that 'Man Must Measure'. But it is true that in a world of elites and masses *managers* must.

7　Schooling

'LOVE, WORK AND KNOWLEDGE,' Wilhelm Reich wrote in 1945, 'are the well-springs of our life. They should also govern it.' But where are love and life in school? And what has school knowledge to do with work? Reich, a revolutionary psychoanalyst persecuted under the Nazi regime and shunned by progressives in Germany and later in the USA, where he emigrated, had his own answers to questions like these. Taking Reich's ideal of self-development through shared work as the touchstone, in this chapter we look at the force of 'commonsense' conservative views in relation to the way that work in classrooms is actually organised. Progressive ideas in education, stressing child-centred learning, have been much-publicised. But conservative and reactionary ideas now dominate in official education policy, forming part of an ideological battle of great importance, particularly to science and scientific leaders.

In his Presidential address to the Royal Society in 1978, Lord Todd chose to defend scientific education in the face of 'ideology'. He spoke of:

the tightening grip of the State on secondary education which has

---

HEADPIECE The eagle's chicks. (Drawing by William Steig from *Listen Little Man!* by Wilhelm Reich.)

been a controversial feature of recent years. Beginning with a laudable intention of ensuring that every child should have an equal opportunity, some of our political masters now seem bent on imposing uniformity and pushing eglitarianism to the point of ignoring differences in ability, and opposing any ideal of selection or segregation on merit grounds. In practice this means that education is to be organised and run in accord with one political ideology.

This was a political speech, from a scientist to scientists and, through the press, to a general public. Particularly interesting is Todd's stance against 'uniformity'. This is a powerfully value-loaded term, which evokes as its opposite 'freedom', something which of course we all leap to defend. We should not leap too quickly, however, because Todd uses the term 'uniformity' to mean 'levelling', and what he is arguing for is using the school system to maintain the differences between elites and everyone else. His criticism of uniformity seems heavily ironical when we note how uniformity shapes work and school life, and that science education is perhaps the most oppressively uniform of all. To interpret ideological debates about education and to think about alternatives in education it is necessary to look quite carefully at what goes on in classrooms, as a form of work and in relation to work generally.

The starting point of many radical critics of education, and of the abuse of science through maleducation, is a belief that 'An education system reflects the purposes of the society it serves.' In one sense there can be no doubt about this, as the present chapter will argue. School work is organised in many ways that are identical with the patterns of industrial work, and school knowledge is organised on the hierarchical pattern of the 'expert' knowledge that such societies run on. Nevertheless the case concerning education is not open and shut. In the late eighteenth century, the formative years of the industrial revolution, 'Dissenting Academies' formed part of an alternative culture for groups excluded from the ruling institutions. So did the Workers' Educational Association and Ruskin College, Oxford, at a later stage. During the last fifty years progressive ideas in education have been an active part of culture and have left as a legacy an apparatus of open-plan schools and 'resource centres' with which even conservative educationists have to work. Educational institutions are perhaps among the most complex in industrialised society and present-day institutions, resources and movements carry still-active residues of past radicalism.

What is radical can, however, also pass over into conservatism. During the decades when schooling was being built up, as a device for teaching workers' children the Three R's, science itself involved reading and writing of a different sort; reading 'the book of nature' and, through the expanding power of industry, writing whole new books of 'nature' – the world of resources and artefacts, techniques and institutions that we live in today. While working-class education was being developed as a system, science was moving as an active project of middle-class workers giving new meanings to the world. Today most of those meanings are themselves conventional, a new common sense, and scientists themselves have become workers. Today, in school, science figures massively as product, received consensual knowledge, and not as process. The active transformation of nature and society and ideas which science signified two hundred years ago has passed over, with the systematisation of knowledge and of schooling, into the transmission of facts and techniques and attitudes to facts and techniques. Science in school today mediates no active reading and writing of the world. As an explicit component of the curriculum it can often be quite marginal. Yet paradoxical as it may seem, because of the massive presence of 'bodies' of knowledge modelled on Science (the *product* of successful sciences) all reading and writing in school is hostage to the pattern of established science.

This is why Lord Todd's argument as a spokesman of the establishment against 'uniformity' is so disingenuous. The Labour Party's policy is an ideology, but Lord Todd's plan for 'unbiased advice' from the Royal Society's elite members is not? If the argument for equality of opportunity is 'laudable' (Lord Todd's condescending term), then how are we to value the domination within education of models of knowledge deriving from established science? Is equality to mean only equality of access to culture as doled out in schools and not equality of expression within schools for cultures other than the authoritarian culture of science? These critical responses to a conservative view of education and science should be kept in mind as classroom work is examined.

CLASSROOM WORK

Knowledge, says the school, is out there; teachers' business is to get it in there; and kids' business is to take it in. But a lot of what passes for learning in classrooms is simply alienated reproduction, in which people work on material foreign to their experience. One researcher on language in the classroom, Douglas Barnes, calls the knowledge which is reproduced 'school knowledge': 'School knowledge is the knowledge which someone else presents to us. We partly grasp it . . . but it remains someone else's, not ours.' Science, of course, is the area of specialisation where this 'body' of knowledge, existing objectively outside the classroom and the school, is most highly developed and most radically removed from everyday experience. Scientific knowledge is 'school knowledge' par excellence. The writing we do in science is often copying whole chunks from the blackboard or out of the textbook. In science, and more in maths, we don't even use *words* much of the time. I know an English teacher who can always tell when the kids have come from Maths.

They haven't spoken for an hour – a whole hour! – and they're exploding with talk.

During the 1960s and 70s some teachers were critical of this state of affairs as it affected the lives of working-class children (especially bright working-class children) in school. They saw that school texts and teachers' ways of talking and presenting issues were often foreign to the ways of talking and kinds of knowledge which the children brought to school with them. More than that, the circumstances in which pupils could talk and write were controlled by teachers so that the 'learners' were unable to adopt their own strategies for coming to terms with material and ideas and other people.

Writing

This showed up particularly in writing. It's useful to refer to some research published in the mid-70s by the Schools Council. The researchers distinguished between the three main modes of writing, and studied how these were distributed across subjects and ages in the work of 11–18 year-olds in school. The fundamental mode of writing, out of which the others grow, was called 'expressive'. In this the writer speaks personally, the assumption being that the reader has an interest in what the writer sees and thinks. A second mode, 'poetic', originates not so much from a person as from an experience, so that it is meaning rather than truth or falsity which is at stake in writing, and empathy which is called for in reading. The third mode – and this is dominant, across ages and across subjects – was called 'transactional'. The implicit claim in this mode is that what is being said does not rely on the support of a particular person or experience, but rather stands on its own as a true statement, and the reader does not expect to have to allow for individual circumstances.

The same research also looked at the audience which seemed to be assumed in a given piece of classroom writing. Six divisions were made: writers writing for their own benefit (as in notes or a diary), for an adult whom they know and trust, for the teacher in a kind of interchange, for the teacher as an examiner, for friends, for a wider public not personally known. Already with 11-year-olds, and overwhelmingly among the oldest pupils, it seemed to be assumed that writing was done for examiners to read. In summary, the research indicated that school writing discriminates against personal knowledge and is dominated by the need to give the 'right' answers. The research results are summarised in Tables 1 and 2.

Within the dominant mode, transactional writing, there are subdivisions. Broadly, we can distinguish between writing that conveys information and writing that seeks to influence readers by persuading or by laying down rules. More narrowly, within the 'information' category, the researchers distinguished between degrees of distancing from actual events. The researchers' categories of 'distance'

132

were, recording (saying what is happening), reporting (what happened), narrating or describing in a less 'photographic' way, beginning to generalise, generalising to the extent of forming a classification, speculating on things which might happen in other circumstances and, finally, theorising in a controlled way about what makes things happen the way they do. The research indicated that at least until the minimum school-leaving age most writing was in the first five categories, 'Rarely was there any sense that they were taking part in a dialogue in which new ideas could be aired and explored.' The pupils were learning to write in the way that school text books 'talk', and were writing for readers who, they knew, already had the answers. This is the learning of school knowledge.

WRITING IN CLASS

TABLE 1

Percentages of year-group sample

TYPE OF WRITING	YEAR 1	YEAR 3	YEAR 5	YEAR 7
Transactional	54	57	62	84
Expressive	6	6	5	4
Poetic	17	23	24	7
Miscellaneous	23	14	9	5
SENSE OF AUDIENCE				
Self	0	0	0	0
Trusted adult	2	3	2	1
Pupil-teacher dialogue	51	45	36	19
Teacher as examiner	40	45	52	61
Friends	0	0	0	0
Public	0	1	5	6
Miscellaneous)	7	6	5	13

TABLE 2

Percentage of subject sample

TYPE OF WRITING	English	History	Geography	RE	Science
Transactional	34	88	88	57	92
Expressive	11	0	0	11	0
Poetic	39	2	0	12	0
Miscellaneous	26	10	12	20	8
SENSE OF AUDIENCE					
Self	0	0	0	0	0
Trusted adult	5	0	0	4	0
Pupil-teacher dialogue	65	17	13	64	7
Teacher as examiner	18	69	81	22	87
Friends	0	0	0	0	0
Public	6	0	0	0	0
Miscellaneous	6	14	6	10	6

Data: Nancy Martin and others, *Writing and Learning Across the Curriculum 11–16*, Schools Council, 1976, pp. 21 and 27.

SAMPLES OF CLASSROOM WRITING

From an expressive discussion of the human body's blood defences:

And while in one part of the body this terrible mammoth struggle for life is going on, the rest of the blood calmly continues its normal life, fetching, carrying, cleansing, and purifying, just getting on with it. It makes me feel uplifted almost, as if it's some kind of Pilgrim's Progress allegory: O Death, where is thy sting? O Germ, where is thy victory? I can almost *feel* the blood at work in my own body. It's as if I could focus down to see these extraordinary things happening a million times a minute all over my body. How odd.

The same object can be responded to in different ways. The next account is moving towards the poetic, communicating an experience which triggered the imagination:

Turn the wood over and a new animal appears no longer gentle and docile but a snake with a large staring eye. Its body has been cut in half and only the head section is left. It has pouched cheeks and a smooth body. It stands upright with its head held high in the air. All it needs is a tongue.

Another account of the same object moves into the transactional, giving details in a quite powerful photographic way with only a little emotive coloration:

A rough piece of wood although smooth in places 6–8" long. Well grained especially along curves. Ugly looking knots which look soft although they are hard. Worn into the shape of a camels head or from another angle a dog bone. 1 L bend. Coloured from cream shaded to a dark brown dry and brittle from one end it could appear to be a fossil (nose end).

Zig zag cracks appear across the grains folation like marks under neck.

Here is transactional writing at its rock-bottom informational level, correct but dry as a bone:

The human skeleton consists of two groups of bones. The *vertebral column*, *spine* and *skull* and the limb girdles and limbs. The upper limb girdle is called the *Pectoral girdle* and the lower is called the *Pelvic girdle*.

And here is transactional writing at a high level, moving into theoretical speculation:

To me the fact that the colours in the sequence kept returning in the same order suggested that as the polythene was stretched it was going through a regular pattern, and I thought this might be due to the atoms in their long chains (for polythene being a polymer is made up of long chains of atoms) slipping over each other.

The majority of writing in school is like the skeleton piece – low-level, transactional, teacher-as-examiner. The knowledge it carries may be useful to the writer but unless he actually becomes a medical worker this bare catechism is not *his* knowledge, it is just words on loan. The 'blood' writer, however – an eighteen-year-old – is making the knowledge work for her. The second driftwood sample is too inwardly directed to be accepted in most classroom contexts as a 'piece of work', and in being polished for public consumption it would shift towards the skeleton style, losing the 'poetry' feel of its bare perceptions. The first driftwood sample wouldn't be generally acceptable in science lessons, nor would the blood 'story', though it poses real questions about the relationship between personal identity and scientific knowledge. The last piece is real science, in the sense of theoretical discovery, but it's not 'properly' scientific because 'I' is at the centre of it.

(Samples taken from Carol Burgess and others, *Understanding Children Writing*, Penguin, 1973.)

TALKING IN CLASS

'What is striking is the very small amount of *individual attention* of what might be called the Plowden-type, the teacher can give to pupils in a 'typical' teaching session. The 2.3 per cent recorded ... amounts to only one minute and twenty-three seconds in a one-hour session. This is clearly very small, even with the addition of the fifty-four seconds attention every hour ... which the pupil receives as a member of a group; an approach specifically recommended... in the Plowden Report as a substitute for the more favoured total individualisation of the teaching-learning process.' (p. 61)

'The distinction between the teacher's and the pupil's interaction patterns are clearly apparent. The great bulk of the teacher's contacts are with *individual* pupils. By contrast, the bulk of the pupil's interaction with the teacher is as a member of the whole class.'

TABLE 3

TEACHER INTERACTS WITH	Pupils' record	Teacher's record
Individuals	14.6	71.6
Groups	9.4	9.4
Whole class	75.9	19.3
	100.0	100.0

PUPIL RECORD	Task related	Non-task	Total
No interaction	40.5	25.1	65.6
Interacts with teacher	12.4	3.4	15.8
Interacts with pupils	5.2	13.4	18.6
Total	58.1	41.9	100.0

NATURE OF TEACHER QUESTIONS	% of all Q's	% of all obs
Task: fact	29.2	3.5
closed questions	18.3	2.2
open questions	5.0	0.6
Task supervision	32.5	3.9
Routine	15.0	1.8
	100.0	12.0

NATURE OF TEACHER STATEMENTS	% of all stmts	% of all obs	
Task: fact	15.4	6.9 ⎫	
ideas, problems	5.6	2.5 ⎬	9.4
Task supervision			
telling what to do	28.1	12.6 ⎫	
praising work or effort	2.5	1.0 ⎬	23.2
feedback on work/ effort	21.4	9.6 ⎭	
Routine			
information	14.5	6.5 ⎫	
feedback	4.5	2.0 ⎬	12.1
critical control	5.1	2.3 ⎪	
small talk	2.9	1.3 ⎭	
	100·0	44·7	

TABLE 3 – continued

OTHER TEACHER INTERACTIONS	%	% of all obs
Gesturing	8.5	1.9
Showing	11.7	2.6
Marking	45.3	10.1
Waiting	8.5	1.9
Story	4.0	0.9
Reading	15.2	3.4
Not observed or coded	2.2	0.5
	100.0	22.3

Data: Maurice Galton, Brian Simon and Paul Croll, *Inside the Primary Classroom*, Routledge and Kegal Paul, 1980.

Results from a study of children and teachers in 58 classes in 19 schools in Britain, ages ranging from 8+ to 10+.

Power Across the Curriculum: Routinisation, Standardisation, Fragmentation

The critique of writing and talking in school – 'language across the curriculum' – is profound and highlights many conflicts within both the structure of knowledge, e.g. between literature and science, and the structure of the education system, e.g. between departments in a school, or between the primary and secondary sectors of schooling. Yet the critique is to some extent misdirected and, for this reason, part of the problem of ideology in education. This issue is not, at root, one of language. What is at stake is power; and a powerful use of language signifies a profound sense of *place*. Critics of 'school knowledge' are generally aware of this and show it by their concern over working-class children's experience of school as a place foreign to their needs and identity. They often do not go on to connect the facts of school life and language with the place of school work in work in an industrialised society. School knowledge is not, in any special way, *school* knowledge. It is an aspect of standardised, abstracted and professionally policed knowledge, and as such it is one dominant form of social control in a 'modern' society.

The essential similarity of school work and 'work' (industrialised work, wage work) is the focus of a critique of school knowledge which goes beyond the school boundary into society generally. Like the world of work, school work is patterned by routinisation, by standardisation and by fragmentation.

The classic managerial strategy in industry has been to control work by mechanising or automating labour processes, thus determining the sequence and form of workers' living labour. As a classroom parallel consider the role of textbook exercises like comprehension tests in English and computational exercises in maths and science. Materials of this kind are used (more by some teachers than others, and more in some subjects than others) as a means of controlling classes by task-attention. Rule by Workcard is a deservedly notorious form of this practice, for the work card edged its way into the repertory of teaching materials under the guise of 'individualised learning materials'. What actually happens, much of the time, is that *isolated* question-answering becomes the mode of classroom work. Table 3 shows how dominant teachers are even in primary schools, supposedly the home of child-centred practice. 'Child centredness' became a new form of teacher-directedness in many teachers' practice. This is routinisation, rather than mechanisation proper. That will have to wait until microcomputers become widely used as 'teaching aids'. Then classroom managers can go all the way.

Teaching from mechanical materials like this is clearly related to the predominance in classroom work of the more mundane forms within transactional writing. It is possible for a text book to pose 'Describe what happens when fertiliser is used on long-stemmed varieties of corn' as an answerable question. But 'Discuss what is important about the Green Revolution' can only be asked and answered through dialogue with persons. Teaching materials can pace and police routine reproductive work, but not synthetic or analytical thinking. A text can act as a 'person', as powerful works of literature testify. But kids in school meet very few books as friends. Most texts are there as models for reproduction. This is why the preface to poet Adrian Mitchell's latest book forbids its use as a school text.

As shown by the research on classroom writers' sense of audience, it is not only the materials worked with but also the anticipated use of the product that determines how work is done. Control over the form of products is called, in industry, standardisation. Standardisation means that each worker's work can be subjected to quality control on the basis of a limited number of measurable qualities such as size, quantity, and various formal properties of shape. So too with much writing for the teacher-as-examiner. 'Too brief', 'Where was your last lesson's work?', corrections of spelling and punctuation errors, are the stock in trade of the hard pressed teacher-as-quality-control-inspector. Not reading for meaning or for exchange of personal insights, but reading for correctness. And hence, once innocence dies, writing for marking and not for expression or evocation of a human response.

The industrialisation of schooling bites most deeply, perhaps, in fragmentation. In an industrial process, operations are separated

137

from one another so that the final product is invisible to the individual worker, whose work on each part is paced by the speed of the process as a whole, either mechanically by a moving track or personally by first-line supervisors and progress chasers. In schooling, curricula are devised 'out there' along with exam schedules, thus pacing and policing the labour of both teachers and pupils. Within the classroom the teacher functions as a first-line supervisor within a division of labour which is externally imposed through academic specialisms and academic institutions. In manual work, fragmentation; in intellectual work, specialisation; and in school work, learning chopped into 'lessons' by the ringing of a bell, isolated tasks emerging from a book or teacher, 'subjects' whose boundaries are marked by walls and doors. All of a piece.

Putting the Con into Concepts
Overall, look at the relationship between thought and work. First, in industry: during modernisation different performances are demanded of workers, and 'craft' know-how of workers becomes transformed via scientific studies into feed-and-speed tables and optimal process conditions, managers' definitions of work groups and systems analysts' definitions of what systems are for. Once this takeover of know-how has happened the whole map of knowledge can be redrawn by 'thinkworkers' in a way which shows nothing to those who by now are paid to be non-think workers, nothing except suddenly re-routed roads, signposts, and tighter traffic regulations. With less of themselves required by the work (this is called 'dequalification' of jobs) workers are forced to settle for less in immediate job content.

Depending on how much learning and unlearning takes place in the job, workers eventually lose the skills they don't get paid to use, and as 'deskilled' workers recruited to dequalified jobs they ultimately settle for less in the labour market too. This process, the central process of 'modernisation' of work, is a process of preconceptualisation: workers' work is scientifically thought out, workers (non-think workers) are not paid to think, and anything they can do beyond what is required in production becomes systematically devalued.

In school, whether by teachers' design, neglect or impotence, classroom work often closes the same circle of degeneration. If learning is personal development, as progressive ideology claims, how is it to be produced except by working on the personal materials of feelings, experiences and insights, through expressive talk and writing? As the map of development is redrawn for the 'learner' by others who are experts in their fields of child psychology, curriculum planning, and educational technology, the stakes are raised to an intolerable level. Those who follow their own maps are labelled deviant. Those who accept being labelled as ignorant have to prove

138

otherwise at the passing-out exam. The pupil's non-school knowledge is devalued, no matter how street-wise they may be.

Preconceptualisation is a political strategy in knowledge, and the thread which draws together work and school. The hidden curriculum of working and schooling is a struggle for meaning and control, between working people or working kids and experts (or 'professionals'. Most teachers wouldn't claim the other title – though why they should want *this* one the Devil alone knows). It is a genuine struggle, and positions in it are not easy to find.

Many teachers feel they can act as guides but are uneasy about their roles as police and managers. In some primary schools and secondary schools' English departments the value of talk as the context for writing, and the world as a context for learning, is upheld in the face of pressures towards mechanical transmission of information or testable 'basic skills' and academically abstract knowledge. Nevertheless the battles are often lost and in some places the values may have been long submerged in the 'professionalism' of class control (Table 3 again). In any case it is an unequal struggle when cast, as it often is in practice, as a struggle between individuals and institutions. No teacher is free from the pressure to 'cover the ground', by Cook's tour or forced march if necessary. Despite the fragmentariness or apparent absence of political consciousness in education and despite Lord Todd's elitist claims to be above the struggle, the institutions of abstract knowledge are a major site of political struggle. What this means is something which needs to be explored further.

KNOWLEDGES AND JOBS

Quite rigid divisions (more rigid in the USA than in England) are made between 'academic' and 'practical' subjects. The former are taken by those who will become college students (they do sciences and maths for example) while the latter, dismissively called 'shop' in America, are the future mechanics and technicians. The mechanic may take engineering and woodwork classes and learn to harden tool steel without knowing anything about crystal structure, while the college chemist picks up no knowledge of practical metallurgy. They occupy different niches in the division of labour, and were fitted for them by the subjects' niches in the school timetable. The academic tends, in fact, to drive out the practical, so that subjects like art, needlework, and craft or 'design and technology' in its up-market version, become top heavy with written work and abstract knowledge. I know a school where three-fifths of 'needlework' time is spent writing!

It is very tempting to see this as a simple reflection of industrial demands for skills, supported and entrenched in school curricula by governments committed to industrial productivity as a main criterion of social progress. It is that, but it is not simple. It is also a process of teachers in down-market subjects wanting to upgrade their image – this is called 'professionalisation'. But this is itself part of a larger process, of selecting for social roles, inculcating attitudes and identities, and legitimating a social order, a process which operates through knowledge.

In many industrial jobs it is techniques rather than capabilities which are desired, and techniques could in lots of cases be taught on the job. Employers and their shareholders would, of course, prefer the State to pay for training, so that technically inappropriate knowledge is often accepted by industry just because it comes 'free'. This is also why there is so much British interest in plans for an Open Tech with an 'industry-led curriculum'. If students go on to 'higher' education – that is, more abstract learning – it is seen to be more efficient if they arrive with an infrastructure of facts, theories and techniques supplied by school, however inappropriate this might be to the needs of those pupils who suffer the lessons and the exams but do not go on to university or a polytechnic. School science carries out the future scientist's induction into the hierarchy of knowledge, though of course it is set very firmly in the context of explanation and not the context of discovery. 'Discovery' methods notwithstanding, it is the logical reconstructed product of sciences which is presented, so that you can't see the joins too easily, while the piecemeal, sometimes erroneous, always personal and heavily institutionalised nature of scientists' work is magicked out of consciousness.

School knowledge is functional for future employers and future teachers if only in cutting their overheads. Whether it in fact produces useful workers depends on the job. But as we have noted above, schools and school knowledge have more purposes than this. For example, however ridiculous exam results may be as an indicator of performance in the job, at a time of high unemployment the grading of school leavers by exam results serves as a ready-made means of partitioning the labour market into employables and unemployables. Academically inflated standards are current in many sectors of the labour market, which means that the majority of candidates are always 'underqualified' and if necessary, the entry requirements can always be raised. . . The device is also used for matching college application levels to numbers of places. Exam results may be no indicator of what a person knows or can use. They are, however, part of a market mechanism, in that they serve to legitimise both a level of unemployment and a hierarchy in formal education (from limited retraining in 'skill centres' through 'techs' and polys to university level) matched to the hierarchy of jobs.

It Isn't What You Do It's the Way That You Do It
The ideological functions of school knowledge go deeper in personal

terms. As well as legitimating the workings of the labour market it also serves as a framework for reproducing varying social identities. Consider 'the lads' (the term is from Paul Willis' book, *Learning to Labour*), working-class children who are deeply aware of the distance between themselves and the middle-class setting of schooling. Because it is held in thrall by a systematically unreal curriculum and the dominance of transactional forms of languaging, the humanism of many teachers appears effete and ineffective, and this gives the lads an image of intellectuality against which to define themselves as anti-intellectuals. And for the 'ear 'oles' – the children who accept the definitions and purposes offered by the institution and its knowledges – school is grasped as a ladder of individual advancement. They climb it until they are pushed off as most people must be, because in a hierarchical society ladders get narrower as you go up; they then nurse a resentment against the values which they dutifully swallowed – neat, impersonal, law-like, modest knowledge. School knowledge is a con because it promises success to those who follow the rules, without showing that the rules change for those at the top. The 'real' science, of Nobel Prize winners and 'Great Men', is not the rule-bound, book-led, exam-ended labour of the classroom. School knowledge directs its protégés into 'normal' science. Whether they play by the rules or defy them, whether they are ear 'oles or deviants, pupils are being set up as pieces in a game which has vicious historical moments. When anti-intellectual manual workers and deferential mental workers find themselves unemployable or threatened by unemployment their logical political response is fascism, the authoritarian mode which promises to make the trains run on time, to elevate 'natural' feeling over reasoned reflection and, above all, to enforce the rules.

There is also another, related, deep structure in school knowledge, which operates not in terms of perceptions of social roles but in terms of workers' understandings of the nature of knowledge. There is work that you occasionally do for fun and there is 'the syllabus', which is the work that you do for exams. Exam work has value essentially as a way of getting that piece of paper, which is meant to be some kind of career ticket, and represents the first step in accumulating career capital if you are going to be in an occupation where 'careers' are the norm. Much of the knowledge to be consumed and regurgitated in exam-passing is abstract and no more useful than jigsaw puzzles. This is true of the maths taught to secondary school kids. Less than two per cent will ever need to use, say, matrix mathematics, and they will in any case get a computer to do the computations for them. So why don't maths departments teach maths as pure game? Because school work is supposed to be work and you don't enjoy doing real work. Exam work is transactional in the basest sense. It is done for exchange: I have done this 'work', now give me a piece of paper to certify the labour expended.

Any usefulness, for most children most of the time, is an incidental bonus.

This notion of knowledge as a commodity is at its purest and most explicit in IQ testing. IQ testing is totally removed from any context in which real knowledge, what Douglas Barnes calls 'action knowledge' in contrast with 'school knowledge', might operate. A special room, a special text which you haven't seen before, no conferring, a time limit which, like the whole situation, is non-negotiable, and a set of abstract questions whose 'purity' of content is supposed to distil out from you the fundamental components of your ability and know-how. The situation is unreal, the questions are unreal. But the results are supposed to mean something. What they mean is that test devisers, and those who take note of test results, imagine that knowledge and know-how are 'things' which exist independently of contexts, and therefore can be measured in one situation and assumed to stay the same in all others. But the ability to score high on an IQ test will not mend you a broken bicycle, or tell you how to stop the local council from putting a motorway through your back garden, or get you a Nobel Prize, or help you get the household back together when you've all just had a flaming row. If you score high on IQ tests you score high on IQ tests, and THAT IS THAT. Intelligence, knowledge and know-how are something else. They are use values, and their use value is that they can enable you to do different things in different situations. Commodity-form knowledge of the sort that helps to pass IQ tests and exams has very little to do with how to live, except as a first gold-brick of career capital.

School creates attitudes to work, less by talking them at pupils than by presenting them as the 'natural' form that work takes. The fundamental relations of production as routinisation, mechanisation, fragmentation, commodity production, are built into knowledge and identified with the experience of learning and 'self-development' in school. The inevitability of wage work is constructed in school by the same kinds of relationship between worker and work that later police the wage bargain in 'real' work: preconceptualisation and commodity production. It would be profoundly radical to challenge these forms of organisation in school, just as it would in 'work'.

THE POLITICS OF SCHOOLING

As they have emerged in this discussion the politics of schooling are not essentially a matter of ideology and counter ideology. What people say, and what people try to pass off as authoritative knowledge, does matter. But it only really matters when those views are supported by experience, and map on to the actual organisation of practical life, so that those views are experienced to be true, or to at least have some truth in them. There is, for example, a kind of sense in Lord Todd's arguments for a system of education which

142

produces an elite of scientific specialists and accepts the fragmentation of school knowledge. Science as a form of work today *is* carried out and run by elites. And just because the 'body' of knowledge is fragmented it is able to reproduce a characteristic narrowness and hard intensity of outlook which is one of the aspects of science most widely valued.

This does not mean that Lord Todd's truth, partial and conservative as it is, has to remain even partially true, because the actual organisation of practices can change historically. But to think beyond constraining truths we need a notion of what we are up against which goes beyond 'ideology'. Ideology is ideas, but ideas linked in specific ways to specific practices in society with their own distinct interests. Lord Todd puts forward ideological propositions. Clearly, so does this book. But to get beyond the limits of any particular ideological understanding we need to understand the practices in which it figures. That is, we need to look at the way that ideas map on to and grow from practices, but without getting bogged down in the confrontation of ideologies, in debates. An alternative philosophy of education needs this kind of understanding of schooling. It needs to go beyond ideology and counter-ideology, to understand and to counter *hegemony*.

Hegemony is the name for a total relationship between the practices of different groups in a society, some of which are dominant and others subordinate, some at the centre, some at the margins. The groups that are dominant do not remain so by the open use of force. This is called repression, not hegemony. Short of repression, dominant groups stay dominant by virtue of subordinate or marginal groups' failure to find alternative ways of working or thinking. Our ways of living and learning are ordered by a definite, discernible apparatus. People who do know what to do, people who don't, places for being taught in, places for working in, books for leisure, books for school; these landmarks of culture are outcroppings of the underlying structure of hegemony.

Love, Work and Knowledge

Identifying these divisions does not itself solve any practical problems. But it at least gives us a bearing from where the chapter started, with Wilhelm Reich's question. Where are love and life in school? What has school to do with work? And where, for that matter, does knowledge lie in schoolwork? The answer to this last question is not so obvious as it might have seemed, for knowledge is both central and peripheral. Knowledge is central in the structure of power which relates experts and the non-expert people whom they ostensibly serve. For the professionals, knowledge really is power albeit of a limited kind. For pupils in the classroom, however, knowledge is peripheral. Abstract, detached, low-level transactional (commodity-form) school knowledge is peripheral to their experience and their needs, whether or not they are going to end up in college. All learners need to relate knowledge to themselves through expressive working, in talk and writing and acting. They need to explore, in 'poetic' form how experiences and visions can be made more real and objective by sharing them with others. All this is heavily discriminated against in the forms of languaging that are sanctioned in school, and the sanctions are licensed by the power and success of 'scientific' ways of languaging reality – modified for general consumption because the worker bees can't be fed royal jelly.

Schooling reduces and routinises the ways in which we experience, think and live. It does this in the same sorts of ways as sciences reduce and routinise the content of work generally, subordinating personal development to an external 'body' of purposes and understandings. But the culture of science is only one way of reading and writing reality. There is no saving clause in this bad contract between people and knowledges, the products of intellectual work. There is no good reason why other ways of reading and writing reality could not be just as real, active and valued.

The question of work in school has been the central strand of this argument. School has a lot to do with work because it introduces workers to the conditions of industrial labour; yet it also has too little to do with work. Teachers often do not find it meaningful, in terms of what they understand by 'education', to allow real work into the classroom. They tend to prefer easily controlled, abstract models.

What, finally, about love and life? If work and knowledge could govern school life in the way that is being hinted at, then life in general would become part of what happens in school, and the Three R's would have to dissolve themselves in reading and writing reality. Love, if we mean caring actions in relation to other people, would be nearer to an everyday reality than its present existence in teenage comics and romantic novels. But where is the place of love and life in a monetarist economy? And what do experts and professionals do to keep them in their place? This is not a textbook question. . . .

Schooling a Scientist

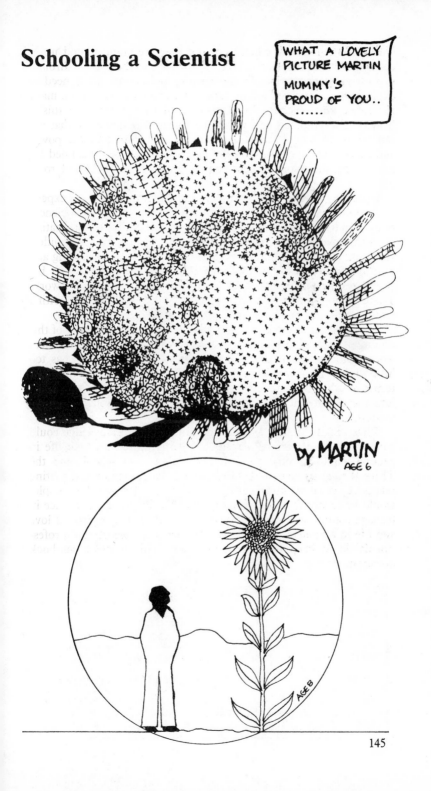

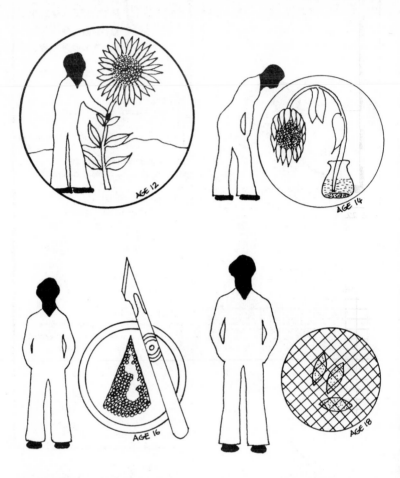

MARTIN M.			*luntine lecture*	AGE 20
BIOLOGY II		WK 16		C/A271
9.50 penalty	9.00–10.30	11.00–12.30	PRACTICALS 2.00–5.00	*private with P.*
MON	BIO ᴹᴷ	CH ᴹᴸ	BIO (LAB 16)	ESSAY Nº 2
TUES	CH ᴴᴸ	BOT ˢᴾ	BOT (LAB 21)	" "
WED	BOT ˢᴾ	BIO ᴹᴷ	BOT (LAB 19)	READ P.H.6.
THUR	BIO ᴹᴷ	CH ᴺᴸ	CH (LAB AT)	W/U P/Notes
FRI	BOT ᴬᴮ	BIO ᴬᴮ	BIO (LAB 16)	W/U P/Notes
SAT	BIO ᴸᵂ	—	SHOPS –(Party)	W/L P/Notes

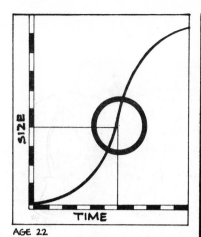

AGE 22

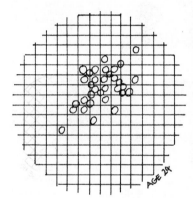

AGE 24

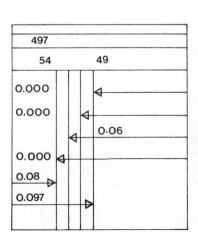

AGE 26

MARTIN WHY
DID YOU MARRY ME!
YOU'RE ALL WORK
WORK, WORK
THERE'S JUST NO
PLEASURE IN LIFE.

8 A Whole History of Inventions

'I AM CONFIDENT that from the same weight of steam-engine machinery, we are now obtaining at least 50 per cent more duty or work,' wrote the inventor of the steam hammer, James Nasmyth, to a colleague in 1852. 'The modern steam-engine of 100 horsepower is capable of being driven at a much greater force than formerly, arising from improvements in its construction, the capacity and construction of the boilers, etc.' Nasmyth's steam hammer impressed Karl Marx. It 'works with an ordinary hammer head, but of such a weight that even Thor himself could not wield it,' Marx noted, marvelling that a hammer weighing six tons with a fall of seven feet could crush a block of granite to powder and yet could tap a nail into a piece of soft wood. The controlled power of machinery was the wonder of the day.

Wonder at technology is of a different order today. Great feats of civil and mechanical engineering have given way to minute feats of low-power electronic wizardry, and the machines of the microelectronic revolution work in small silent ways inside boxes whose outside appearance need betray in no way the chips' presence or function – washing machines, phones, typewriters. Visible spectacle lies within the bounds of the small screen, in the power of computer graphics to generate video images of startling reality or strangeness. If we step back, however, from the physical scale of the machines,

---

HEADPIECE Photograph by Henri Cartier-Bresson, from *Man and Machine.*

some of the same aspects begin to be visible across time. Of steam engines, for example, Nasmyth remarked, 'Although the same number of hands are employed in proportion to the horse-power as at former periods, there are fewer hands employed in proportion to the machinery.' New generations of power machinery were enlarging the ratio of capital to labour employed in an industry so that there were increasingly fewer hands per machine. Today this has a name – two names: technological redundancy and structural unemployment.

Machines and Combinations
Machines were designed to take over human tools and wield them on a super-human scale. Nasmyth again:

> The characteristic feature of our modern mechanical improvements, is the introduction of self-acting tool machinery. What every mechanical workman has now to do, and what every boy can do, is not to work himself but to superintend the beautiful labour of the machine. The whole class of workmen that depend exclusively on their skill, is now done away with. Formerly, I employed four boys to every mechanic. Thanks to these new mechanical combinations I have reduced the number of grown-up men from 1,500 to 750. The result was a considerable increase in my profits.

Apart from increasing profits, one purpose in the development of these mechanical combinations was the putting-down of workmen's combinations – that is, trade unions. The quotation above is taken from an 1868 report of a commissioners' inquiry into the organisation and rules of trade unions and other associations of workers. The machinery which Nasmyth is describing was installed to break the power of skilled engineers who had held widespread and long-lasting strikes in 1851.

If Nasmyth is quite matter of fact about the political use value of machinery, Andrew Ure, in his 1835 *Philosophy of Manufactures*, positively crowed about it. Writing of a warp-dressing machine, introduced to break a strike, he said, 'The combined malcontents, who fancied themselves impregnably entrenched behind the old lines of division of labour, found their flanks turned and their defences rendered useless by the new mechanical tactics, and were obliged to surrender at discretion.' The general principle of technology – that is, science and technology – was perfectly clear to the politically sensitive Ure, 'When capital enlists science into her service, the refractory hand of labour will always be taught docility.'

Machines are machines. Technologies are something more. They are 'mechanical combinations', further combined with – and introduced and invented in the context of – established ways of working and entrenched social and class interests.

The first ribs of Joseph Paxton's Crystal Palace roof being hauled into place by a gang of longshoremen on 5 December 1850, ten weeks after the first column was erected. Built to house the 1851 Great Exhibition (in which Nasmyth's press was an exhibit), the Palace itself was a technological spectacle and a demonstration of the effectiveness of modular construction and standardised design, presaging an era of design-based mass production and consumption.

We might define a technology as the use value of a machine or system of machines; changes in technology are intertwined with and give direction to changes in the division of labour. The warp dressing machine introduced a new division of labour which removed the old locus of workers' power. Nasmyth's self-acting tool machinery supplanted grown-up men by boys. More powerful machinery, in the absence of corresponding increases in markets for products, supplants workers in work and places them in 'the reserve army' of the unemployed. In moral terms nineteenth-century defenders of the political role of technology got into ridiculous tangles. Ure claimed that while progress in machinery displaced adults 'thus rendering their number superabundant' it also 'augments the demand for the labour of children and increases the rate of their wages'. But at the same time, it was felt necessary to keep children's wages down, because otherwise parents who could not find employment at a living wage would send their children out to work at too tender an age. Morality was the icing on the cake. Its coherence did not matter much, so long as entrepreneurs had effective power over the composition and wage rates of the workforce. This the technology gave them.

Seeing science as work, a history of technology needs to be tackled as a history of transformations in the forces of production, in which machines appear as a main part – but only a part – of the cultural apparatus of a society. The industrial revolution marks the beginning of a continuing series – a long revolution – of transformations in the forces of production. Water power then steam, factories (called 'manufactories') then mechanisation, electric power, the moving assembly line, the internal combustion engine, automatic control, robots and satellites, bugs and spares, windmills and nodules. In the course of these transformations the composition and distribution of the workforce has been through many changes, in terms of sex, age, colour, nationality, national location and the actual skills and experience required. The whole process has been – and is – one of struggle, over market advantages and capital accumulation, over health and standards of living, over authority and the rights of tradition, over skills and the identities of working people which are rooted in and acted out through them. Seen in this framework a technology and its historical transformation maps an interface in cultural struggle. A technology shows at any point in time the balance of forces between capital and labour, and between various sections of workers – men and boys, women and men, whites and blacks. This is true also of relations between workers in metropolitan countries and workers at the imperialist periphery. The challenge and the difficulty of historical studies of technology is to read back from the machines – the brute facts of development – to the cultural and political processes, the tacit but ultimately no less brutal and material facts of struggle.

151

Citing Ure and Nasmyth, Karl Marx proposed that it would be possible to write a whole history of inventions since 1830 made for the sole purpose of providing capital with weapons against working-class revolt. 1830 is a convenient date to signify when technology in its modern sense began to emerge as a significant component of the forces of production, and the manufacturing classes began to consolidate their new hold on power in an industrialising society. Today we might follow Marx's suggestion in looking, for example, at the tachograph ('the spy in the cab' of longdistance haulage workers) or the IBM 3750 computer, which can monitor and log work performance rates on 'electronic mail' systems, destinations and content of phone calls (barring selected numbers from access) and all movements of staff through pass-coded areas with computer-operated door locks. In the remainder of this chapter we shall focus on just two inventions which show some of the complex relations between capital and labour and experts and workers: electronic type composing in the print industry, and the injectable contraceptive, Depo Provera.

THE PRINT OF CAPITAL: MICROELECTRONICS IN TYPESETTING

Until very recently the compositor was a 'labour aristocrat', highly skilled, well paid, in control of the conditions and pace of work. George Dodds described him well in 1858:

> He has to manage the punctuation, which authors too generally care very little about; and he often rectifies an occasional error arising from haste . . . His eyes guide his fingers (or his fingers guide themselves) to the cells where the proper letter-types are to be found; and the formation of letters into words, words into lines, lines into columns, columns into pages, and pages into forms or sheet-surfaces, takes all his powers – mental, visual and digital. He also has to 'mind his p's and q's', not only in the literal sense of that phrase, as the p appears on the type like a q to the unpractised eye, but in many a figurative sense also.

When the types were locked up into formes for printing a whole newspaper sheet, the resulting object was both heavy and awkward, so that a degree of physical strength was needed, too. On this basis of skill, hand-compositors (or 'comps') were able to build a strong resistance to employers' efforts to mechanise and speed up the process of type setting.

In particular they were very sensitive about the 'demon of boy labour', and justly so given how effectively jobs in printing (as distinct from the setting of type) were redesigned through technological development so that unskilled, young, low-paid workers could occupy them. Compositors' response was to enforce a strict control on entry into the trade through long apprenticeship. Entry

Applegarth's vertical rotary printing machine at *The Times*. Mammoth machines of this kind, fed by eight boys on top and off-loaded by eight boys beneath, were used from 1848 and attained speeds of up to 12,000 impressions an hour. (Illustration: Lardner, *The Great Exhibition and London in 1851*, 1852)

into jobs was controlled too, through the patronage of 'chapels' (the print unions) which in turn operated through families, fathers securing jobs for sons from generation to generation. This patriarchal culture was one of the bases, alongside the control of apprenticeship and of the actual process of setting type, of the power of the men's trade organisations.

The advent of the popular press, with its mass readership, in the mid-eighteenth century, had entrenched the trade. But as early as the end of the nineteenth century speeds of printing had been so greatly increased by steam-driven rotary presses, that hand composition was a serious bottleneck in the production of print, particularly since apprenticeship procedures were succeeding in restricting the flow of labour into the trade. Hand composition was also a bottleneck in production of profit because compositors, as wage workers, were pricey.

Each letter was printed from a separate piece of type, and so at the end of the print run, when the formes were broken down, there was a vast amount of type to be redistributed or 'dissed' back into the type cases. When the comps successfully resisted attempts to use women for dissing, employers sought to do away with it altogether. The Linotype machine was invented in these circumstances, and it did away with dissing. Each line-of-type was cast afresh from matrices held in the Linotype machine. Matrices were immediately machine-sorted for re-use without requiring a subsequent stage and after printing the slugs of type were simply re-melted and fed back as liquid type metal for recasting. Other kinds of hot-metal typesetting machine followed, notably the Monotype.

Operating these machines was no longer heavy work but the job was still highly paid. Women now wanted to become hot-metal typesetters. In Edinburgh in 1872, when the male comps were on strike, a small group of women established themselves. Women were rapidly replacing male clerks in offices as the typewriter began to be employed for producing business correspondence, and in fact Monotype machines had the same 'lay' as typewriters, whereas Linotype keyboards had a special lay which only trained compositors knew their way around. Male compositors dealt with the general situation by defining hot-metal setting as a part, but only a part, of the total craft of compositing. They insisted on using the Linotype. They also insisted on the right to maintain the machines and to do every job associated with their use, even down to sweeping the floors. In this way the trade was kept whole and women were kept out. The assault on women in the Scottish print industry early in the twentieth century emphasised many 'caring' moral issues such as the dangers of lead to sensitive constitutions – which clearly men could bear but not women – and the damage which might be done to women's child-bearing faculties by heavy work. The campaign was successful and the trade remained all male. Even through the two world wars, when women were able to enter heavy occupations like agriculture and engineering work in munitions, the men kept their hold on entry to the trade, their control of work rates, and their high levels of pay.

Enter the Computer

Computer controlled phototypesetting is a radical break from hand or hot-metal type setting where the actual 'setting' is done by physically moving reversed solid shapes and fixing them into a frame. In photosetting, images of individual characters are formed optically

154

19th century book: hand composed page, printed by steam driven flatbed press, set by apprentice trained male compositor in 70 minutes, made-up forme weighing 26lb transported to press room by labourer.

20th century book: computer photo-set page, printed by computer driven offset press, set by retrained female ex-copy typist in 8 min, image 'recipe' transmitted to imposition room on floppy disk weighing 25 grams.

at around three thousand characters a second, either flashed up on a TV screen (a 'VDU', Visual Display Unit) or by using a laser beam on photographic film. The process moves on from there to produce either photographic positive prints known as 'bromides', which are physically pasted up to make an image of the whole page, or to a further electronic composing stage in which the ghosts of 'typeset' items are marshalled into complete pages by operators sitting at computer keyboards in front of VDUs. In the latter case the characters never appear in physical form until the final page image is sent to the platemaking process and transferred to a metal printing plate.

The output is not all that is radically changed. The input keyboard does not drive a mechanical system producing slugs of type, but a computer which stores the strings of characters in a memory unit. The high computing power of microprocessors means that computers for typesetting can be made to perform most of the calculation and checking which was once the skilled compositor's preserve. 'Justifying' (that is, setting lines to a fixed length as on this page), breaking words at the end of lines and checking spelling using dictionaries can all be built into the machine, so that the inputter's job is just that – keying in a string of words. The material can be viewed and edited, though not necessarily by the same person who did the input, on a VDU then passed electronically to the photo-setting stage.

The transformation of the machine, and also the tranformation of the technology, acts very broadly upon the composition of work, implying a recomposition of jobs and of relationships between different kinds of workers. A 'smart' machine of any kind, one which can make many of the necessary choices in performing a task, creates simple work requiring fewer abilities than did pre-smart machines. Typesetting became dequalified through smart machines, requiring less training and less extensive experience. But simple work, or rather simplified work, implies also some rather complex work somewhere else in the division of labour, analysing the tasks, writing the software to fulfil the necessary functions, designing the hardware. Although the immediate job of the worker in typesetting now is simpler, the reproduction of the whole technology – workers, machines, software – is greatly expanded because now the functions of design and maintenance have passed to quite new sections of the workforce. Electronics designers, systems analysts, programmers, electronics engineers, are all implicated in typesetting work now, along with the institutions of training that their occupations require. Although it is not directly visible in the smaller job of the compositor, the extent of the socialisation of production and reproduction has greatly increased. More different kinds of workers in more places are now implicated in the activities of the worker at the input keyboard, and the dequalification and marginalisation of one position in the division of labour – the typesetter's – implies a more general recomposition of relations between sectors of workers. At least some (in this case probably the systems analysts) have become more central and maybe more powerful, though the power is not necessarily visible to them. It may just be that the 'power' appears only in the labour market as one section of workers, the typesetters, lose out in job security and employability while another section, systems analysts and programmers, find themselves in greater demand. This is not really power that can be wielded in a positive way.

Men's Work?
Compositors' control in the new technology is very much less. They can no longer service the machines because electronic components are designed to be unserviceable and replacements come through an external supplier. The social life at the keyboard is shattered. At the old keyboards comps were so fluent that they could talk to each other as they worked, the text simply flowing through their optical channels and out at their fingers on to the machine. The new keyboards are standard typewriter-lay keyboards, and the men are much more awkward with these. In fact any good office typist could set type faster than most of the men using the new machines. The old machines required space which the new do not, so that work is now done at workstations closer together, set in rows. 'Battery hens'

the men call themselves. The whole feel of the work has altered and the identity of the workers is undercut. The keyboard is 'female' like an ordinary typewriter, the relationship with the machine is 'female' in that the men can no longer service it, and the work situation is 'female', like a typing pool.

The new technology has in fact provided the opportunity for women to enter the trade. Over the past decade the increasing use of electronic communications media and the wide distribution of decentralised print workshops like the local 'copy shop' or in-house print room, which does typesetting on an electronic memory-type-writer and prints on an offset-litho machine, has outflanked the traditional apprenticeship controls of the men's unions. Independent and necessarily non-union women typesetters (because of the union's resistance to apprenticeship of women) worked, often in their own homes, using machines such as IBM Varitype electric typewriters with golf-ball print heads and small electronic memories. The print unions have reacted to this situation by blacking typography coming from non-union shops, arguing that in doing so they were helping to combat the sweating of outworkers' labour. Women typesetters experienced the response as an attempt to resist the upgrading of their skills as typists into 'proper' print industry work. In the longer run it seems clear that women are now accepted as here to stay by the unions, which are now amalgamating with unskilled workers' unions and recruiting women, but often into a separate section – or ghetto – within the union. The struggle between men and women workers over the status and occupancy of typesetting jobs, far from being over, will work itself out further within the union in terms of demarcation agreements, gradings and so on.

In relation to the employers, the men adopt the now-traditional comps' stance in the face of new technology, insisting that training on the new machines be seen as part of training for the whole range of typesetting, paste-up, editing and associated tasks. This is partly to keep control of entry into the area (now hardly a 'trade' in its old sense) and partly to discriminate against typists who, although far more competent as keyboard operators, have little experience of the other components of composing-room work. The apprenticeship system has been completely undercut. Early in the century lads were apprenticed for seven years, attending classes, working, and earning little. When they came out of their time as skilled men, which was what they could now officially call themselves, they resented any attempts to have 'their' work done by other workers who had not made the same sacrifices. By 1975 apprenticeship was cut back to four years, and in the 1980s, partly to accommodate the reskilling of existing members, unions have agreed to a modular, flexible system of courses which make it possible in principle for new entrants to become 'skilled' in easy stages. Women presumably will take care to take the full range of courses in order to guard against any future declarations of second class citizenship.

Skills and Technology

Typesetting in the context of computer technology is deskilled. It is not, however, quite so simple as that. 'Skilled' men, that is, apprentice-trained union members, are no longer called upon to use the range of experience, dexterity and intellectual qualities that were an intrinsic part of their work as hot-metal setters. Their position in the division of labour has been dequalified, they have been deprived of the location for exercising their full range of accustomed abilities, and they have therefore been deskilled. Faced with this attack on the traditional bases of their organisation and job security they have had little option but resistance to the new technology. For women and men outside the union the situation is different. The typewriter lay of the new machines and the simpler job content of keyboarding mean that the threshold level of ability and experience for 'real' typesetting is now much lower, making access to print more direct and access to jobs in the print more feasible. For them the new technology creates the possibility of upgrading their skills and pay, from typing and 'clerical' wages to composing and 'skilled' wages.

The capstan lathe, which many women were trained to operate in World War II, is a machine designed to deskill metal turning in mass production. Tools are advanced in a set sequence by tripping the capstan lever and take a depth of cut which is preset by a 'skilled' setter.

The issues are even more complex than this; it is not just a matter of having skills, more or less of them. Skill is not a thing, and the question of what is defined as skill and what does not count is central. Even at the level of the physical content of work matters are complicated. Work in printing was traditionally too heavy to be women's work. But can all men lift more than all women? The keyboard on the Linotype is big and wide and manly. Can the sizes and shapes and layouts of work materials not be changed? Why not? The pitch of a screw thread determined the amount of effort required to tighten a nut, so if nuts are 'too hard' to tighten why not redesign them? Units of load – fertiliser sacks, bags of cement, hay bales – are interesting. From men's point of view, on one hand, they should be as big as possible, so that men can claim the work as theirs and exclude women. On the other hand, if the loads get too heavy men will damage themselves and weaken their job-control relationship with employers who require the largest possible loads shifted in each journey to maximise profit. The actual sizes of bags and bales map quite visibly the relative strength of male workers on the two fronts of masculinity and patriarchal power, and resistance to capital's squeezing-out of profit. Ergonomics and time study are the 'science' of drawing this kind of line, that is, of rationalising (on employer's terms) the current state of political forces in the workplace by adjudicating on the units of workload.

It is entrepreneurs and their agents who traditionally have the power to introduce new kinds of machinery into workplaces. In this way, and through the preconceptualisation of tasks which is designed into the machines, the owners of capital establish their hegemony in work, partly through changes in technology. Within this frame male workers, not just in the print trades, have been able to negotiate a subordinate form of power which entrenches their position of superiority relative to women. This backfires when men's political organisation is weak, because the depressed wages of women – a result of male hegemony – make women workers an attractive alternative for capital and thus creates the conditions for employers' attempts to introduce female labour to replace male. But generally men have been able to define the conditions of work in such a way as to dequalify women – being 'too arduous', requiring too much formal education, demanding a certain temperament which just happens to be 'typically male', and so on.

Microelectronic technology in general gives women a second argument to bolster the obvious one. The obvious argument is: We *can* do the work, be plumbers, roofers, car mechanics, truckers, skilled turners. Many women can. But there is a counter argument citing physical differences between the average women's and men's bodies, and the cultural differences that there happen to be between the experiences that women and men have as they grow up. The men may have the maths and science that women mostly don't. The

159

sacks of cement may be too heavy for the woman builder to lift comfortably. This is not the end of the line. She may get help, she may go to a supplier of different size sacks, she may deflate the male egos of men who near-rupture themselves to prove how manly they are. But still it is an obstacle. Low-power, small scale, 'smart' machinery like microelectronics-based machinery takes out the need for the muscle stuff and much of the long-experience stuff, so women can say much more forcefully: there is *no* reason why this should be men's work exclusively. We have right of access too.

In general anybody might do any task if the technology were designed to facilitate it. Physically impaired people can be mobile and dextrous if they have the right machines. Less strong people can lift heavy things if they have a hoist or can get others' cooperation or can change the demands of the task so that smaller loads will do. Unskilled people can do skilled work if 'smart' machines are designed to make that possible.

Capital and Deskilling

There are two traps lying in wait of the unwary assumption that labour-saving or skill-enhancing machinery might imply a utopia just around the corner. There are two traps – and both of them are capital. First trap: design and implementation of new machinery takes time, especially when it is radically new machinery. Time is money, and if money is capital then it won't be spent unless there is the strong likelihood of a significant return on investment. Who will invest in cheap mobility devices for physically impaired people? Who will invest in machines to make work lighter, or accessible to people without much experience of abstract reasoning in maths or the sciences? Unfortunately the answer to that is obvious, and it is the second trap.

Employers will invest in any new technology which does these things if the net effect is 'deskilling'. To define what this means we have to make a careful and non-obvious distinction, between simplifications and changes which make a task lighter or more accessible to more people on equal terms, and those which operate through market power to increase the advantages of one group of workers relative to another; women relative to men, boys relative to men, unskilled men relative to skilled men. In the latter case employers will invest because through the new technology they may get cheaper labour, more control over the product, more control over the intensity or duration or quality of work and so, more profit. If it is paid for by capital, labour saving is not *labour* saving, it is *cost* cutting for the capitalist. If it is paid for by capital, one worker's skill enhancement is another worker's skill devaluation, with an overall net saving of money. This is what capital *is*.

The idea of alternative technologies and alternative products is very important and powerful. There are always alternatives to the way things are done now. But thinking about alternatives has to be broader than just drawing-board dreaming, about machines. It must also encompass the social relations within which the machines will be used – technology is the *use value* of machines. The social relations of capital make deskilling, somewhere in the recomposed workforce, an inevitable aspect of new technologies. New technologies without deskilling require ways of linking people together in society other than through markets and money.

DEPO PROVERA: BIRTH CONTROL AS A FIX

Birth control technology is big business. Nearly half the mothers in Africa using birth control are on Depo Provera (DP). Thousands use it in Asia and Latin America. Ten thousand women, mostly Polynesian and Maori, were given the drug in New Zealand trials.

Somewhere between three and five million women in the world are probably on DP. It has been used by – or on – more than eleven million women since it was introduced by the pharmaceuticals multinational, Upjohn, as an injectable contraceptive marketed especially in the Third World. Partly because of its own characteristics in use, and partly because First World feminists, learning from the Pill, are fighting shy of this new 'wonder' technique, the drug has become the centre of fierce controversy.

The American Food and Drug Administration rejected Upjohn's 1978 application to market Depo Provera in the US because of suspected cancer risks. In 1982 the advisory Committee on Safety of Medicines recommended that it should be licensed for extended use in Britain but they were overruled by the Minister of Health on the grounds that it could be given to mentally handicapped girls without their consent. DP is recommended and distributed around the world by the IPPF (International Planned Parenthood Federation) whose medical director argued that the British Minister's decision was dangerous because it might inhibit the use of DP in countries where many women die in childbirth and many die young. The viewpoints and interests are intensely contradictory.

The Use of DP: Upjohn, the Multinational

Contraceptives in the Third World constitute a massive market, and one which can be exploited conveniently through the State because of national birth control policies. Medical workers in the Third World are often paid on commission to secure the 'acceptance' of drugs by women, and Upjohn has paid well over four million dollars in bribes to hospital workers and state officials in order to secure the marketing of its product. Depo Provera's use value to the manufacturer is the same as that of any commodity – it can be manufactured and marketed at a profit. The particular characteristics of this technology which give the drug its unique use value in Upjohn's product range stem from the fact that the drug is injected at long intervals rather than taken daily by mouth, and from the population control policies of Third World states.

The Use of DP: The State

Depo Provera is ideal because of its efficiency in both physiological and economic terms. One injection guarantees contraceptive action for months; three months is the nominal period but, to cover themselves against statistical variation, the manufacturers' dosage remains in a woman's body for anything up to a year. There is no daily routine, no chance of forgetting, guaranteed infertility. This means that there need be no careful education in the drug's use, as required by the Pill, so that medical workers' time is conserved. The same advantage accrues from the fact that insertion of the 'fix' is fast and simple, in contrast to the diaphragm or IUD (intra-uterine device). If cost efficiency is the prime criterion, as it is with State planning programmes, then the health and comfort of the consumers of the technology is subordinate.

Cost efficiency in administration and profit efficiency for the manufacturer combine to make follow-up and monitoring something to be avoided if possible. If screening and follow-up were generally practised then the conveyor belt value of the technology would be lost (exactly analogous with the health and safety situation in factory production), DP's cash value would drop and so also its deployment by planners. If pelvic examinations and cervical smear tests were carried out fewer women would continue to accept the technology when there was a less distressing option, that is, a more fully tested and less hazardous method of contraception needing little clinical follow-up.

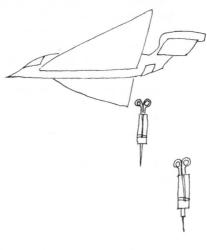

(Cartoon: Merrily Harper)

162

In international terms of funding and political control, birth control is a non-exclusive alternative to increased food production, and in this connection Robert McNamara of the World Bank has compared 'miracle' drugs favourably with 'miracle' grains. 'Family planning programmes are less costly than conventional development projects,' and they can yield very high economic returns. The contraceptive fix, Depo Provera, fits within the political spectrum of other technical fixes employed in imperialist politics and links the interests of metropolitan powers like the US, multinational companies like Upjohn to the governing elites of peripheral States in the Third World.

The Use of DP: Professionals

DP is easy to prescribe and administer. It can be deployed effectively without the 'patient' actively cooperating and therefore can operate across cultural barriers of all kinds. Thus, for example, the majority of the ten-thousand New Zealand women in the Depo Provera trials were not given interpreters' assistance. A common way for doctors in Britain or America to justify their 'professional' distance, that is, their habitual manipulation of female patients is to argue that 'these women' wouldn't understand anyway. 'These women' are generally black or Asian, the doctors generally white and male, and medical practice frequently racist and rooted in male domination. Depo Provera technology, with the passive role that it prescribes for the woman patient, provides an ideal medium for professionals' intervention in women's reproductive cycles in these conditions.

Because it is so economical of professional's time, as well as of State money, DP is particularly 'desirable' for medical practitioners who treat illiterate women or women who, for other reasons, are difficult to police as users of a contraceptive technology. Mentally ill patients, 'promiscuous' girls in care, women coming out of prison and young girls undergoing abortions have all been dosed with DP without their consent, sometimes under anaesthetic, sometimes under a pretence that the injection was something else, sometimes without any explanation at all other than that it was a 'necessary' part of medication. In Leeds it is a common practice for Asian women to be given Depo Provera injections routinely, in conjunction with rubella injections following the delivery of a child.

With an efficient fix like DP, professional judgment can become physiological fiat in one swift step. Injectable contraceptives, in this connection, are a violent technology. That is, they facilitate the violation of women's bodies by professional workers with privileged access. An alternative to physical coercion or deception is the use of financial or political sanctions as in the case of black women in the USA who have been threatened with withdrawal of welfare payments unless they 'accepted' injections of Depo Provera. In this respect DP plays the same role in professionals' policing of social

163

standards as do other 'desirable' means of fertility control such as sterilisation and IUDs.

This is not to imply that all use of DP by medical or social-work professionals is in this mode. Medical and social workers may, and many do, consult with the women who choose to be given Depo Provera. There are, however, professionals who function as not-so-soft cops in the reproduction of women's subordination in both First and Third World countries, and for them injectable contraceptives have a distinct use value. It was highly disingenuous of the medical director of the IPPF to say that Depo Provera had been used by eleven million women: it was used on many of them.

The Use of DP: First World Women

One of the cries of protest which went up when the British Minister of Health denied Depo Provera a licence was that the drug was not only used on illiterate and otherwise culturally disabled women. It seems to be increasingly requested by white, middle-class women, sometimes in professional occupations, who could be expected to know what they are asking for. There are three characteristics of the technology which might recommend Depo Provera to them. It contains no oestrogen, only a progestogen, and thus is free of a particular hazard associated with oestrogen in the contraceptive Pill; it is injected quarterly, and therefore relieves the user of daily routines and apparatus; and it often has the eventual effect of causing menstrual bleeding to stop (amenorrhoea).

An Indian woman doctor, a feminist, said of this: 'I am an acceptor of DP and use it myself. I am happy to have amenorrhoea.' It was clear from an interview that she gave (*Spare Rib* 116, March 1982) that she has a thoughtful and considerate approach to the women whom she advises, and in particular that she is wary of the 'fix' mentality of Western medicine. In other cases too, notably also with women doctors, personal care and support may be part of doctors' approach to the use of contraceptives. Nevertheless, as Jill Rakusen points out (in Helen Roberts, *Women, Health and Reproduction*), 'Personal care, support, and reassurance, even concerning apparently non-threatening side effects, is only caring, supportive, or reassuring if it is based on sound evidence. Commonly, this evidence is lacking.' The main reason for this is commercial pressure to get DP on the market and keep it selling. It was mentioned earlier that adequate follow-up and monitoring work has not been done on Depo Provera because the examinations themselves might discourage women from 'accepting' the drug. The list of the drug's 'side' effects is long. Most women have bleeding trouble with DP which in women suffering from malnutrition, frequent in the Third World, adds the burden of anaemia to the inconvenience and despair of the basic condition. Bleeding after childbirth can be aggravated and prolonged. Eventually, after a year or so of use, the opposite

164

disturbance – amenorrhoea – takes over. What is not known and what the manufacturers' research is not designed to determine, is whether or not long-term use of DP will damage a woman's metabolism or hormonal balance and perhaps lead to permanent infertility. Many women find it distressing rather than a convenience. The bleeding increase is sometimes controlled by prescribing women oestrogen. In Chile one expert reckoned that about four in ten of his patients were eventually given oestrogen, which completely undercuts one of the logics for the drug's use. DP affects breast-feeding mothers and their babies absorb large quantities through their mothers' milk. Breast cancer, cervical cancer, liver cancer and endometrial cancer are all indicated, through animal testing, as possible effects of large doses in humans.

Statistically, the whole area is a maze of ineffectual surveys, misleading presentations of data, failures to gather data, and partial concealments of the results that do exist. A special leaflet in Urdu, for example, in an English town with a high Asian population, neglected to mention the severe bleeding which most women suffer, and doctors often fail to grasp or convey the information that headaches, dizziness or amenorrhoea may have anything to do with the drug. Once a woman has been injected with Depo Provera she is committed to suffering whatever effects it may have on her, for up to a year after the injection. Even literate women with access to the literature find the use of DP a gamble rather than a decision.

The Use of DP: Third World Women
Two main characteristics of Depo Provera – it is injected and injections are infrequent – give the drug a distinct use value for many Third World women. Because it is injected there are no visible signs, packets of pills, trailing threads from IUDs, and no surgical treatment. The fact that Western medicine has breached traditional culture to the point where many people believe that medicine from a needle is the best kind, only adds to the perception of DP's use value. Women in South India suffer from anaemia and malnutrition, they are heavily constrained by traditional female roles, and are subject to constant sexual abuse by their husbands who generally refuse to permit any kind of contraception on the presumption that it indicates a woman's intention to commit adultery. Men have been known to drag out an IUD in order to assert their patriarchal power. In these circumstances, Depo Provera offers one of very few degrees of freedom in a woman's life. Her opportunity to choose DP may be one of her few opportunities to choose.

The three-monthly frequency of DP injections brings two sorts of advantages with it. First, because it is easy to administer, DP takes effective fertility control to remote regions where qualified medical staff are few, and this is a real need when women and children may die in childbirth beyond the reach of a skilled midwife or doctor. An injectable contraceptive in the hands of a minimally trained villager can liberate a woman from the fear of further pregnancy, and a village from the burden of extra mouths making demands on marginally cultivatable land. The second point is more complex. In subsistence farming communities time has little meaning, in the sense that it is taken for granted in the industrialised world. Time, along with work discipline, had to be introduced into the experience and the structure of working people's lives during the rise of Western industrial capitalism. In, say, South India the time-economy of industrialised activity still stands beyond the boundaries of everyday life, and the clock is not part of the apparatus of existence. There is not a time and a place for everything, because agricultural time is not like that. Tradition insists on place in social terms and craft terms. Nature insists on time but not in a regulatory way, for there may be room for negotiation. A fertility control technology which presumes routine – like the Pill – will either fall through the spaces of the indigenous culture and fail or begin to work an industrialising transformation on that culture. It depends on the force with which the foreign technology is supported in the immediate context of women's daily life. DP technology is unusual as Western technology, in that it does not carry routinisation as a necessary component, and this may enable it to carry a higher use value in Third World cultures.

Abuse and Double Standards

Out of such disparate and conflicting use values, how do we make sense of the technology? Actually, terms like those that head this section – 'abuse', 'double standards' – are not much use for this. Abuse, and its implied opposite, proper use, are too single minded in the way that they ask to be applied. In fact it is clear that use value for one interest group often contradicts use value for another. The company's profit cuts across and reduces the woman's security in knowledge, the professional's ease of control diminishes the woman's sense of feeling well and in control.

Although the South Indian peasant woman or the North London feminist actively suffers anguish and pain in using the drug, she may still want to use it. Although she resents her treatment at the hands of the doctor or the agent of State 'welfare' she may decide that she wants the injection. Thus, use for the goose is also use for the gander. It does not clarify the situation to call one of the uses 'abuse'. In practice it is one use against another, and it is precisely because the abuse is a *use* to some real and powerful social interest that it becomes dominant.

What we are up against, in this tangle of use values, is hegemony, the exercise of power without visible force. The hegemony of capitalist firms interlocks with that of national ruling elites in Third World countries but the former is dominant. Use values for Indian development are subordinated to use values for Western capitalist firms. This is partly because of capital, meaning simply money. But it is also because of the power that firms in the North hold over the South through preconceptualisation – the power to give knowledges and products a shape which imposes itself in the place where they are used. In its turn, the hegemony of experts and professionals is subordinate to that of corporations and States. But their use values are a fair way up the heap, they have limited power, maintained for them by drug companies and governments in providing the apparatus of their practice: the drugs themselves as 'tools', the money to support their services, the legislation to give their prejudices muscle. Experts and professionals use that power and thus squeeze the autonomy of their 'clients' smaller and smaller. Women's use values come last. If there is any freedom left they can have it.

This systematic organisation of use values, then, is a competitive, hierarchical, violent, discriminatory technology, which is why feminists, such as the organisers of the Campaign Against Depo Provera, are right to resist it. But the argument that is crucial for countering the technology is not that DP removes control more firmly from women and therefore that its use should be banned. It is the technology, not the drug, that removes control, or rather, asserts and reproduces hegemony. The technology in this instance consists of the drug plus the social relations of its design, research, production, marketing, testing, administration and consumption.

All along the line there are uses, right to the very end, in consumption. To remove the particular focus of these relations, Depo Provera, would displace all the various hegemonies into other locations, other drugs, other medical techniques, other economic channels, other forms of political power. It might be that the result would be an enlargement of the autonomy of women in relation to their own control of their fertility or it might not. In these circumstances, lined up against elites and concentrations of capital, with alternative technical fixes in surgery, intra-uterine devices and oral drugs, the deck seems heavily stacked.

The intention of this account of DP is not to generate pessimism or to cool the powerful and necessary anger of women fighting for a right to feel well and be free. But the fight will need to be orchestrated across a range of interests which the notions of use and abuse simply are not adequate to capture.

Like 'abuse', the notion of 'double standards' is also one-dimensional. Use values for South Indian women are qualitatively different from those for North Londoners. First World feminists live in a culture which accepts and approaches history as a potential framework for social progress. For First World women the history is bitterly familiar after the 1970s' experience of the Pill. It is a history that they want to unmake and as in many struggles, the bitterness of frustration spills over into the campaigns. For women in the

167

Third World control over their fertility is a desperate need, but such power as comes within their reach is not self-made power because taking control of their history is not yet part of their history. This means that they make their choices with a different – and necessarily different – consciousness than their First World sisters. In these circumstances, single standards could not possibly exist.

Thanks to the flexibility that their capital gives them, multinational companies exploit the uneven cultural development and uneven economic development of regions of the world to their advantage. If a company's products cannot meet the standards which an industrialised culture insists on then they will try to sell them in countries with different standards. This is not a matter of higher or lower, but of *different* standards, because for there to be a single standard across countries there must be a single culture. This does not excuse the dumping and testing of products not proven safe – and sometimes proven unsafe – in cultures where safety is less a priority than survival. It does not excuse the cynical playing off of one culture's values against those of another when the dominant value – market value – overarches and undercuts both. But what it does do is to demand a fuller recognition of the reality and qualitative differences of culture from those who engage in political struggles around technologies.

THE FORCES OF REPRODUCTION

At the beginning of this chapter it was suggested that a history of inventions needs to be tackled as a history of transformations in the forces of production. Following that, a computer technology was discussed as a process of the socialisation of work, and a contraceptive technology was discussed in terms of multiple and contradictory use values. How might we weave all these together?

A first difficulty is that technologies tend to be understood as essentially connected with machines, which is in fact the connection in which this chapter made its first approach. There are no machines involved in fertility control. The link in describing both of this chapter's examples as technologies lies not in machines, but in the idea of ordered systems of use values which centre in some kind of physical artefact. In the first case it was a type of machine, an electronic machine. In the second case it was a chemical substance, a drug. It might also have been a material such as coal or pre-cast concrete or refined cornstarch. Each has a distinct technology – energy or chemicals, construction, junk food – associated with the use of the material in a given historical frame. The unifying idea in examining any of these would have been the same, that of looking for the social relationships which connect one use with another in such a way that a pattern of uses arises which we acknowledge by naming it 'a technology'.

A second difficulty is mostly terminological. If we are looking for

168

shifts in forces of production, how does this get hold of a history of *reproduction* such as the story of Depo Provera? The notion of 'a force of production' stresses the apparatus of a system of practices – its extent, its internal structure, and its extensive connectedness with other major systems of practice. This notion is as useful in discussing Depo Provera as it is in discussing computerised typesetting technology. Injectable contraceptive technology is geographically extensive, spreading across industrialised and unindustrialised hemispheres of the globe. It has a complex internal structure incorporating many workers in many hierarchical roles, and in central-peripheral relationships with one another and with the consumers of the practices' product. Finally, in being intimately connected with programmes of economic development in Third World countries and with the financial and political institutions which direct those practices, DP technology fills out the picture of a force of production quite fully. The essential emphasis of the notion is on the ways in which practices and developments hinge on things rather than ideas. When we recognise the extent to which Depo Provera technology operates through violation rather than education – the action of things on things – then the idea of reproductive technology as a force of production becomes even clearer.

The Socialisation of Production
Although we did not follow it that far, the socialisation of work implicit in computerised photosetting technology goes far beyond the country in which the machines are used. Microprocessors are themselves products of a widely flung division of labour which spans the American West Coast, the Far East and the industrialised countries in which final products are designed and used. If we had back-tracked through the practices which end up on the other side of the compositors' keyboards we should have covered as much ground, geographically, as Depo Provera manifestly does. The recomposition of work that was noted is only part of an international recomposition of work and skill and unemployment. The struggles over status and employability and control and male/female identity are also only parts of an international but fragmentary struggle.

It was argued that the power in those struggles lay in capital, as money, and in preconceptualisation – the systematic separation of knowledge from the users of products worked up from that knowledge. The recomposition of compositors' work, cultures and identities took place through a capability that modern sciences provide, of designing whole new ways of doing things and making them available to be inserted into another practice which has not generated them. The people who work and live in that practice – in our example, the male compositors – have to accept the new technology or refuse it. Consent does not make the technology theirs, because the working knowledge of it is somewhere else – actually, in *many* places, many knowledges. Refusal opens them to material sanctions such as being locked out or sacked, but also to verbal attacks: they are ignorant (of course they are, that's what preconceptualisation is for), they are Luddites, opponents of scientific progress.

Although we looked at Depo Provera through the notion of use value rather than socialisation, this same kind of process is at work there too. Women, especially women in non industrial cultures, are confronted with new technologies which they did not design and did not ask for in any articulate, collective sense, and are not equipped to appropriate as their own. The power of preconceptualisation – the whole apparatus of research and development, demarcated from consumption – offers only two options: acceptance in ignorance, or refusal. Preconceptualisation in the pharmaceuticals industry produced a technology of fertility control which renders women passive. Preconceptualisation in the microprocessor software industry produced a technology of 'smart' machines which renders the smartest hot metal compositor no more competent than the most typographically ignorant typist.

The Geography of History

We approached DP technology through use values, that is, through each group's ability to appropriate the technology and make it serve some of their purposes. There were multiple and conflicting ways in which this single technology, a single structure within the forces of industrialised production, buttressed or extended or contained the power of people and institutions. It became clear that technology is part of a cultural apparatus, a setting for actions and reactions, almost like a stage set on which actors improvise roles. The doctor administering Depo Provera might be an uncaring, manipulative, agent of a cynical state planning system, or a supportive encourager of women's autonomy within the straitjacket of systematically incomplete knowledge.

Taking up different threads of the action and different props from the apparatus of the technology, groups of actors take up various positions on the terrain. Bribing petty government officials, to smooth the path of the product; vacillating over the safety of the method because the facts are unclear; refusing tetanus vaccinations in Bangladesh because the doctors have previously shot them full of DP and they will not trust the doctors any more. The locations are limited. But they are also distinct, have their own developing plots, and their intrinsic limitations within the general configuration of the playing space. Within those limitations, women improvise around the basic human themes of feeling free, feeling well, knowing what to do and being free to do it. They struggle for whatever freedoms they can create out of the situation. Of course, open up the playing space by adding a new technology – such as free, safe and early abortion – and whole new ranges of action become possible.

Although the geographical dimension of the compositors' struggle was less obvious, the same kind of geography-cum-drama exploration could be worked on that ground too. Multinational microchip innovators, intensely competitive national newspaper operators, managers who want to assert their 'right' to manage, comps who want to hang on to their jobs and status and style, unskilled workers who want to get in on the compositing act – these are the roles, or some of them. The props are provided by the old and the new technologies. Different props lend themselves to different uses in improvisation and because they have introduced some of the props according to a plan some of the actors have also had time to script the action – which is not to say that they can direct it as written.

This is not a fanciful elaboration of the idea that 'all the world's a stage'. It is an attempt to work through the fairly straightforward idea of use value, with its subjective and purposive colouration, and link it to the more mechanical idea of the apparatus of a technology, as a kind of landscape in which use values are created or discovered and put to work. To describe the systematic organisation of the terrain we already have the concepts of commodities, and preconceptualisation. What we are now sketching in is a concept of how groups of people respond to the conditions in which they live and work. Not just to the terrain, but also to the possibilities that they can find in it for realising their images of themselves.

Naming this idea is difficult because so many different kinds of organisation and so many different forms of identity are involved, but a name is needed for future reference. Without meaning to imply that gender, or race, or age, or physical impairment, can be reduced to this one dimension, I'm going to choose the term *class* as the best one to characterise the struggles which go on in this kind of terrain. And the mapping and the actual distribution of such struggles then can be called *class geography*. Class geography is the geography of history. A whole history of inventions is a history of movements in this terrain; and a whole – rather than a divisive – future of inventions would demand that we first rework it.

Taking Apart the Telephone

Divisions of labour in the telephone

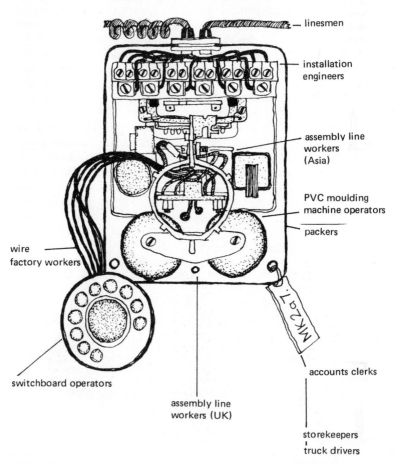

linesmen

installation engineers

assembly line workers (Asia)

PVC moulding machine operators

packers

wire factory workers

switchboard operators

assembly line workers (UK)

accounts clerks

storekeepers
truck drivers

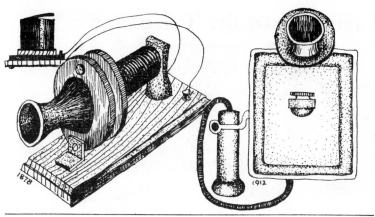

'Modernisation' of telephones and networks is a process of major upheavals in patterns of ownership and work.

173

'a boon'

'a must'

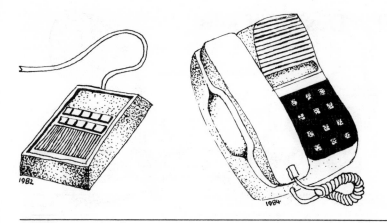

1982

1984

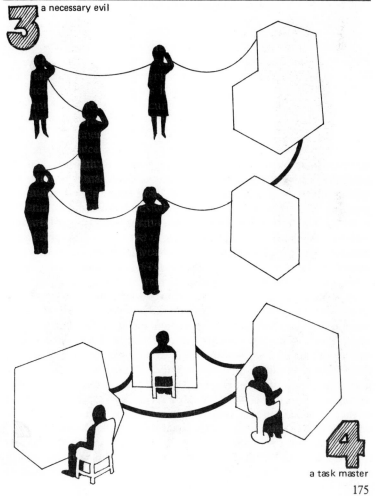

3 a necessary evil

4 a task master

9   Design of jobs

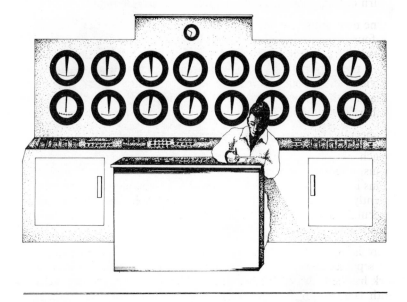

ALL DESIGN IS DESIGN OF JOBS. This may seem implausible but the general basis for the proposition is simple enough. People have the special ability – which sets them off from any other species – to prethink what they do, to create an object in imagination before it is created in fact. What design does, as a systematic and distinct form of work within an industrial society, is to specialise this general human function. This is the fact of industrialised life which is named by the term, 'preconceptualisation'. Preconceptualisation, as a basic social relation of modern science, makes a special form of work which, in turn, fixes what other workers in other jobs must accomplish in producing specified products or services. Washing machines and automobiles are designed and assembly-line workers make them. Catering systems are designed and kitchen workers deliver the actual food within specifications. Education systems are designed and teachers teach within syllabuses and buildings. Science goes into the specification of products and processes and so, through managers, determines work. A more recent historical development is that of science going into the design of the *social* processes in which workers collaborate in the production of products.

The splitting up of work – the division of labour – is not necessarily a scientific business, and only became so towards the end of the nineteenth century. The classic description of the division of

176

labour, Adam Smith's account of pin making in the late eighteenth century, refers to a state of affairs which developed in handicraft production under pressure from the market and from middlemen to turn out more goods at lower prices and more reliably:

> One man draws out the wire, another straights it, a third cuts it, a fourth points it, a fifth grinds it at the top for receiving the head; to make the head requires two or three distinct operations . . . the important business of making a pin is, in this manner, divided into about eighteen distinct operations, which, in some manufactories, are all performed by distinct hands, though in others the same man will sometimes perform two or three.

The benefits lie in far greater output under a single roof:

> Each person . . . making a tenth part of forty-eight thousand pins, might be considered as making four thousand eight hundred pins in a day. But if they had all wrought separately and independently, and without any of them having been educated in this peculiar business, they certainly could not each of them have made twenty, perhaps not one pin in a day.

There is no science in this in the sense that there is no developed and separate analysis of the product and the process. But craft in work has been driven further and further back since the late eighteenth century, as science in design has been advanced, both to increase output and to control workers – to 'teach docility' to 'the refractory hand of labour', as Andrew Ure put it in 1835.

It is a hundred years on from Smith's *Wealth of Nations*, in America in the 1880s and 90s, that we can begin to see the science of job design explicitly developing as a distinct and necessary specialism within the capitalist division of labour. The approach became known as Taylorism, after Frederick W. Taylor who, with his followers, appropriated the term Scientific Management to refer to a system of work measurement and payment.

SCIENTIFIC MANAGEMENT

The Taylor approach rested on a minute, movement-by-movement analysis of each task followed by a derivation of 'the one best way' to carry out each stage, and a subsequent rigid discipline to enforce it in practice. Scientific Management goes far beyond the 'natural' division of labour in manufactories in that each operation within the divided process is closely specified from outside the process itself, by engineers working as agents of management. This was the regime of the man in the buff smock with the stopwatch. The coat might in fact be white or blue, or a suit; and it might have a colour-coded collar or epaulettes depending on the system of rank adopted within a particular company. But the stopwatch was the essential item of dress, and however dressed, the Time and Motion Man became universal – and universally resented.

Taylor sought the 'law' which governed the work of the first-class labourer. Concerning the investigation of a 'first-class man' – a favourite phrase – lifting pig iron, Taylor describes the study carried out by Mr Carl G. Barth 'who is a better mathematician than any of the rest of us':

> In a comparatively short time Mr Barth had discovered the law governing the tiring effect of heavy labour on a first-class man. And it is so simple in its nature that it is truly remarkable that it should not have been discovered and understood years before . . . the law is that for each given pull or push on the man's arms it is possible for the workman to be under load for only a definite percentage of the day. For example, when pig iron is being handled (each pig weighing 92 pounds), a first-class workman can be under load 43 per cent of the day. He must be entirely free from load during 57 per cent of the day. . . . As the load grows lighter the man can remain under load during a larger and larger percentage of the day, until finally a load is reached which a man can carry in his hands all day long without being tired out. When

SELECTING A FIRST CLASS MAN

OUR FIRST STEP was the scientific selection of the workman. In dealing with workmen under this type of management, it is an inflexible rule to talk to and deal with only one man at a time, since each workman has his own special abilities and limitations, and since we are not dealing with men in masses, but are trying to develop each individual man to his highest state of efficiency and prosperity. Our first step was to find the proper workman to begin with. We therefore carefully watched and studied these 75 men for three or four days, at the end of which time we had picked out four men who appeared to be physically able to handle pig iron at the rate of 47 tons per day. A careful study was then made of each of these men. We looked up their history as far back as practicable and thorough inquiries were made as to the character, habits, and the ambition of each of them. Finally we selected one from among the four as the most likely man to start with. He was a little Pennsylvania Dutchman who had been observed to trot back home for a mile or so after his work in the evening, about as fresh as he was when he came trotting down to work in the morning. We found that upon wages of $1.15 a day he had succeeded in buying a small plot of ground, and that he was engaged in putting up the walls of a little house for himself in the morning before starting to work and at night after leaving. He also had the reputation of being exceedingly "close," that is, of placing a very high value on a dollar. As one man whom we talked to about him said, "A penny looks about the size of a cart-wheel to him." This man we will call Schmidt.

The task before us, then, narrowed itself down to getting Schmidt to handle 47 tons of pig iron per day and making him glad to do it. This was done as follows. Schmidt was called out from among the gang of pig-iron handlers and talked to somewhat in this way:

"Schmidt, are you a high-priced man?"

"Vell, I don't know vat you mean."

"Oh yes, you do. What I want to know is whether you are a high-priced man or not."

"Vell, I don't know vat you mean."

"Oh, come now, you answer my questions. What I want to find out is whether you are a high-priced man or one of these cheap fellows here. What I want to find out is whether you want to earn $1.85 a day or whether you are satisfied with $1.15, just the same as all those cheap fellows are getting."

"Did I vant $1.85 a day? Vas dot a high-priced man? Vell, yes, I vas a high-priced man."

"Oh, you're aggravating me. Of course you want $1.85 a day – every one wants it! You know perfectly well that that has very little to do with your being a high-priced man. For goodness' sake answer my questions, and don't waste any more of my time. Now come over here. You see that pile of pig iron?"

"Yes."

"You see that car?"

"Yes."

"Well, if you are a high-priced man, you will load that pig iron on that car to-morrow for $1.85. Now do wake up and answer my question. Tell me whether you are a high-priced man or not."

"Vell – did I got $1.85 for loading dot pig iron on dot car to-morrow?"

"Yes, of course you do, and you get $1.85 for loading a pile like that every day right through the year. That is what a high-priced man does, and you know it just as well as I do."

"Vell, dot's all right. I could load dot pig iron on the car to-morrow for $1.85, and I get it every day, don't I?"

"Certainly you do — certainly you do."

"Vell, den, I vas a high-priced man."

"Now, hold on, hold on. You know just as well as I do that a high-priced man has to do exactly as he's told from morning till night. You have seen this man here before haven't you?"

"No, I never saw him."

"Well, if you are a high-priced man, you will do exactly as this man tells you to-morrow, from morning till night. When he tells you to pick up a pig and walk, you pick it up and you walk and when he tells you to sit down and rest, you sit down. You do that right straight through the day. And what's more, no back talk. Now a high-priced man does just what he's told to do, and no back talk. Do you understand that? When this man tells you to walk, you walk; when he tells you to sit down, you sit down, and you don't talk back at him. Now you come on to work here to-morrow morning and I'll know before night whether you are really a high-priced man or not."

This seems to be rather tough talk. And indeed it would be if applied to an educated mechanic, or even an intelligent laborer. With a man of the mentally sluggish type of Schmidt it is appropriate and not unkind, since it is effective in fixing his attention on the high wages which he wants and away from what, if it were called to his attention, he probably would consider impossibly hard work.

F. W. Taylor, *The Principles of Scientific Management.*

that point has been arrived at this law ceases to be useful as a guide to the labourer's endurance, and some other law must be found which indicates the man's capacity for work.

Taylor's concern with employing first class men did not exactly amount to respect:

Now one of the very first requirements for a man who is fit to handle pig iron as a regular occupation is that he shall be so stupid and phlegmatic that he more nearly resembles in his mental make-up the ox than any other type. The man who is mentally alert and intelligent is for this very reason entirely unsuited to what would be, for him, the grinding monotony of work of this character. Therefore the workman who is best suited to handling pig iron is unable to understand the real science of this class of work.

And to make the point quite clear: no *workman* has the authority to make other men cooperate with him to do faster work. 'It is only through *enforced* standardisation of methods, *enforced* adoption of the best implements and working conditions, and *enforced* cooperation that this faster work can be assured. And the duty of enforcing the adoption of standards and of enforcing this cooperation rests with *management* alone.' (Taylor's emphasis).

The Technocratic Era

With Taylorism and similar lines of development as a fragmenting tendency, dividing work into smaller and smaller parcels nested deeper and deeper within formal levels of study and supervision, some unifying principle was needed in addition to the force of command. Two were forthcoming, within what became known as 'Fordism', after Henry Ford. The first was the moving assembly line. Here was an immediate, objective link and pacing system, to unite the fragmented tasks created by engineers of jobs and to secure the quantity of output in relation to costs of employment, that is, to secure a surplus for capital. In parallel with this was a second principle which operated in the sphere of consumption rather than production. If workers were paid more to work to order, then they would be able to buy themselves into satisfaction and a lifestyle conducive to moderation, since they would then have something to lose other than their chains. This was the philosophical basis of Ford's 'Five Dollar Day'; it was also one of Taylor's principles: 'The great revolution that takes place in the mental attitude of the two parties under scientific management is that both sides take their eyes off the division of the surplus as the all-important matter, and together turn their attention towards increasing the size of the surplus until this surplus becomes so large that it is unnecessary to quarrel over how it shall be divided.' This standard moral argument for 'scientific' approaches to design of jobs was voiced in the context of a prolonged economic depression which in countries like Britain, prompted imperialist expansion. In the USA, where Taylor and Ford developed their systems, it was internal rather than external expansion which was pursued, that is, more intensive exploitation of a rapidly growing labour force with many immigrants. (The ox-like pig iron handler was named Schmidt.)

180

In the politically tense circumstances of depression and within the growing managerial apparatus of manufacturing firms, there developed a dualism of power; those whose eyes were on the market and saw production as a means to more capital, and those whose eyes were on workers and saw everything as a means to more work. Taylor himself was a fanatic about work: 'There is hardly any worse crime to my mind than that of deliberately restricting output; of failing to bring the only things into the world which are of real use to the world, the products of men and the soil.' This placed Taylor in implacable opposition to the labour unions – 'systematic soldiering' was his name for workers' militancy. However harsh his personal views were regarding morality, he and other adherents of the 'works management movement' around the turn of the century believed that science was a means of mediating in the struggle between capital and labour and eventually sublimating its energies. A new elite, engineers and their ilk, would exercise power as neutral agents. Henry Gantt, who worked under Taylor at one stage, stated, 'What we need is not more laws, but more facts, and the whole question will resolve itself.' Henry Ford blew the cover somewhat, since the other side of his 'Five Dollar Day' coin was the infamously brutal Ford police force which broke strikes and heads with equal 'neutrality'. On the other hand, what could be more factual than a blow from a nightstick?

Somewhat less obvious than the stand against workers was the stand taken by proponents of scientific management against capitalists. Once the 'one best way' was discovered, Taylorists regarded it, as well as the 'fair day's pay' that went with it, as a law binding on workers and managers. In political terms scientific management and the wider 'Technocracy' movement which grew out of it were a bid to shift power from the established financier-elites into new elites of technical workers. In the 1920s Antonio Gramsci, the Italian Marxist and Communist leader, noted the responsiveness of Europeans generally to ideas of 'Americanism' and 'Fordism'. What the American vision seemed to promise was an escape from having to accept the pre-war model of ideological conflict and class confrontation. Productivity, expertise and optimisation were technical solutions to political problems. Taylorism appealed to the *produttivismo* of Mussolini and the liberal-capitalist reformers of Weimar Germany alike. But in 1917 the place after America where Scientific Management was most applied was not in Japan, France, Britain or Canada (though the ideas were active here). It was in Russia.

In 1913 Lenin had called Taylorism 'the scientific system of wringing out sweat'. But even then he had noted the 'rationality' that it introduced in an otherwise anarchic system of production. By 1918, after the Bolshevik revolution and confronted by the need for labour discipline if production were to be rescued, Lenin was writing that 'We must organise in Russia the study and systematic teaching of the Taylor system and systematically try it out and adapt it for our own ends.' Later under Stalin, in association with the rise of an elite of proletarian experts, technical and Party cadres, a home-grown variant, Stakhanovism, took over. Named after a devoted proletarian over-fulfiller of production quotas, this approach stressed diligence and effort rather than quantitative formal study. The cult of scientific analysis did not last. It was supplanted by a cult of experts and apparatchiks.

THE ENGINEERING OF TASKS

When it has the power of capital behind it – as it generally does – technology gives managers an effective power over the composition of the workforce and over wage rates. It is not a complete control, but it pre-empts workers' actions because capital buys machines and *then* buys the time of workers to work with them, so that it effectively determines much of the ground on which the actual content of work is negotiated. As opposed to Taylor's belief in 'the one best way', some present day designers of jobs argue that there is 'organisational choice' and that just because a certain set of machines or a certain product market exists this does not mean that job content is inflexible. One general point about machinery in the history of industrial work is that there is more to industry than mechanisation.

The socialisation of production was well advanced before machinery became a dominant aspect of production, that is, the Industrial Revolution was not initially or essentially a technical or engineering phenomenon. Adam Smith's pin makers were engaged in very low-tech handicraft production within an industrial form of work. The big manufacturers of the eighteenth century saw the advantages of social control and reduction of overheads and waste, and of gathering workers together in one large manufactory. Only as a later phase did the same drive for control and profitability lead to a primary place for steam-driven machinery. Benjamin Gott, the first of the big Yorkshire masters in spinning, never used power in his spinning or weaving sheds during a quarter of a century. It was supervision not mechanisation which prompted Peter Stubbs to gather his filemakers from their homes into his works at Warrington.

Of the characteristic aspects of modern industrialised production – hierarchy, fragmentation, routinisation and mechanisation or automation (all encompassed in the notion of preconceptualisation, the pre-thinking of what work shall be done, and how) – hierarchy came and comes first. Thus the London *Spectator* in 1866 praised a workers' cooperative for its adventures in profit-sharing but found one thing lacking:

> Hitherto that principle has been applied in England only by associations of workmen, but the Rochdale experiments, important and successful as they were, were on one or two points

incomplete. They showed that associations of workmen could manage shops, mills, and all forms of industry with success, and they immensely improved the condition of the men . . . but they did not leave a clear place for the masters. That was a defect.

The master came first, with his overlookers who supervised the labour of the hands. Then came fragmentation, the creation of detailed and bounded tasks, a process which may have started in family workplaces under the supervision of putters-out before the factory system took over. With the factory system, routinisation became feasible – the pre-specifying of the form that the work and the detail product should take. It was in parallel with this devolution of tasks that science was harnessed in the development of machinery so that eventually it was possible for masters to translate human detail-labour into a machine process. Mechanisation eventually became the dominant form of organisation, but it was a late development rather than the sole pivot of industrialisation. It is necessary to stress this because quite often progress is identified with machines and machines with efficiency. Neither of these equations holds as a 'law' of economic development.

When Frederick Taylor spent half of his life studying 'The Art of Cutting Metals', in order to apply science to metal lathe work, and when his disciples like Frank and Lillian Gilbreth began to study the layouts of domestic workplaces, machines were at the centre of the scientific study of work. Through such studies, machines were introduced or modified as a means of policing the 'correct', 'efficient' performance of tasks. In machine shops tables of 'feeds and speeds' and a carefully selected range of machine tools were provided as the necessary (and enforced) condition of good, that is, cost efficient work. Ergonomically designed kitchens and kitchen equipment which took into account walking distances, stooping and stretching ranges, weights and measures, served the same function for domestic labour. Linked thus to workers through machinery, Taylorist approaches implied a top-down flow of information, as orders, or as 'education', from experts to working people. Of all the components of preconceptualisation it was hierarchy – *social* hierarchy or authority, rather than functional hierarchy – which was most emphasised.

Information Control
Some later approaches to the design of jobs, especially some 'systems' approaches associated with computers rather than mechanical machines, seem to have a less powerful reliance on visible authority. Among the standard problems in Operational Research are those of allocating resources between a number of demands within a system (coal between power stations, food ingredients between alternative products) or routing a vehicle between a number of destinations or

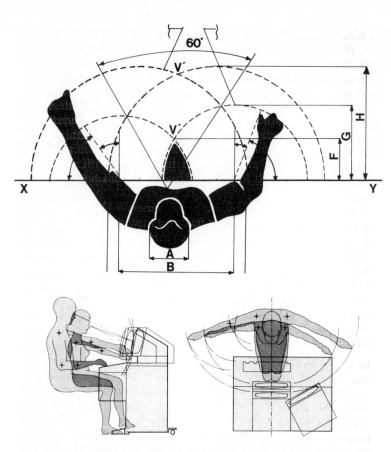

WORK MATERIAL	CUTTING SPEED in m/min								FEED RATE in mm/rev.	
	HIGH SPEED STEEL				CARBIDE		STELLITE			
	Turn		Ream and Thread	Drill	Turn		Turn			
	Rough	Finish			Rough	Finish	Rough	Finish	Rough	Finish
Mild Steel	40	66	7½–15	30	90	180	50	75	0·625/2·0	0·125/0·75
Cast Steel	15	24	3½	12	45	100	24	33	0·5/1·25	0·125/0·5
Stainless Steel	15	18	3	12	27	45	22	25	0·5/1·0	0·075/0·175
Grey Cast Iron	18	27	3½	13	60	100	33	45	0·4/2·5	0·2/1·0
Aluminium	90	150	15	72	240	360	120	180	0·1/0·5	0·075/0·25
Brass	75	100	18	60	180	270	90	150	0·375/2·0	0·2/1·25
Phosphor Bronze	18	36	4½	13	120	180	30	50	0·375/0·75	0·125/0·5

'Scientific' management in the Taylorist vein fits workers into work stations designed to reproduce the one best way of working. Conversely, workers have been fitted to the demands of standardised tasks by, for example, recruiting only women seamstresses with arms and fingers of a given length. Taylor's lifetime research on The Art of Cutting Metals was forerunner to present day feeds-and-speeds tables, published in manuals of workshop standards from decade to decade.

184

designing a policy for servicing queues of customers at a service point. However, none of these definitions of problems or the standard techniques for 'solving' them, implies a particular relationship on the ground between experts and workers in the system. Sophisticated mathematical methods now exist for deriving 'best' solutions as cost functions but the philosophy of this more recent approach (since World War II) is less intrinsically authoritarian than Taylor's. Rather than imperatives, transmitted downwards through the hierarchy, 'systems' approaches use the hierarchy as a filter for flow of information – upwards as well as downwards. The main aim is for information to be in the right place so that whoever works there (manager or worker) is constrained to do the 'right' thing in terms of the overall objectives of the system, simply by the limited information which is available.

OR and other management sciences have a different machine ethic from Scientific Management. They treat organisations as wholes, to the extent to which their remits from higher management allow them to, and analyse them as feedback-controlled machines. This information-processing view of control explains why there is a strong connection between these more recent scientific approaches to the control of work and the use of computers. The metaphysic is well illustrated by a view from the early 1960s:

> The development of modern society has much in common with biological evolution. Modern organisations, whether they be industrial concerns, social structures or state activities, are at the present time going through the stage of developing their controlling mechanisms; this they must do if they are to remain viable. Electronic computers and other types of equipment as yet unborn may provide the machinery; cybernetics may provide the theory; Operational Research may provide the techniques for understanding what goes on and why and what decisions and controls are needed.

The newer and older approaches are by no means incompatible. Management sciences with a 'systems' approach may determine the 'best' objectives of the component parts of a system (departments, or particular groups of workers in management or production) and may design the information-controls which will make the system 'naturally' tend towards achieving these definitions. Meanwhile the authoritarian apparatus of Scientific Management may swing into action to turn the objectives into actual definitions of tasks, and to turn the abstract information design into an actual routine of form-filling, reporting and supervision, a telephone system, a mail system, and so on. The OR techniques which deal with actual material flows rather than information – allocation of coal, routing of lorries, splitting of feedstocks – fall somewhere between the two extremes. The techniques do not actually work right down at the level of workplace detail (workers' movements, times, distances) but their mathematical power to calculate maximum or minimum values of economic functions like profit or wastage tends to make them the natural ally of managers whose style is 'one best way', do it as I say.

Social Engineering

Taylor himself was a controversial figure, and his system aroused some strong opposition. In 1915 the US Congress forbade the use of some of his pay and time study methods in military establishments, following trouble at Watertown Arsenal. Nevertheless a broadly Tayloristic approach was general during the 1920s. In the course of a series of experiments on lighting, which looked for the 'one best way' of illuminating a workplace at the Western Electric Company of Chicago, industrial engineers discovered that the workers responded not only to the lighting conditions but also to the attention of the investigators. Elton Mayo, a professor of business administration at Harvard was consulted and there started an experimental programme which ran over twelve years (1927–39), out of which there emerged a line of thinking later named the 'Human Relations' approach. Mayo 'discovered' that Taylorism left a human residue which needed to be mopped up by other techniques addressed to the social environment of the workplace. This resulted in a consultative style of management, welfare provisions and so on. These palliatives are the administrative responsibility of personnel departments rather than work study or production engineering departments, although work study practitioners often need the insights to ease themselves into workplaces and to sugar the pill of work measurement and effort rating. Scientific Management and Human Relations work in harness, each coping with what the other leaves out.

A number of researchers, notably Frederick Herzberg, whose ideas became popular during the 1960s, later came to feel that jobs could be altered from the form prescribed by Taylorism. That is, rather than each worker performing a single narrowly specified task, they might do a job made up of a number of such tasks (job enlargement) or they might alternate between such jobs (job rotation). Although Herzberg himself argued for something more generous than this, the common interpretation of 'job enrichment' has been the recomposition of boring routinised jobs into larger, boring, routinised jobs.

Such approaches were pulled together and reworked during the 1960s and 70s by social scientists working with a 'systems' approach to work. These 'socio-technical systems' theorists saw a happy convergence between technological and social development in history. Technology, they judged, was driving history along just as, in their judgement, it had done since the start of the Industrial Revolution.

186

Automation in industries like chemicals manufacture was producing a new type of worker whose role was not to work in the process but to stand outside and oversee the general conditions within it. Luckily, culture was delivering the appropriate attitudes of 'Participation and control, personal freedom and initiative'. With this lucky convergence of culture and technology, these theorists argued, 'The advance of technology itself has reversed the world of Frederick Taylor.' All workers were soon going to be managers.

Clearly this is technocratic fantasising. The thing to note, however, is the different style of science which these latter day technocrats bring to the design of jobs. Rather than being deterministic like Taylor's 'one best way', they regard the technical and the social sub-systems of work to be mutually interdependent and to have a certain amount of flexibility. As they put it, there is 'organisational choice'. Machines can be used in various ways to produce the same product. Different sets of attitudes in the workforce will enable the changes to be made. The machinery may even be changed if, for instance, that allows a social control mechanism to operate effectively in maintaining quality or raising output. Workers can be organised in small 'autonomous' groups. Machinery can be modified or designed to feed better or 'relevant' information to the workgroups.

As science, the approach is similar to that of OR as outlined earlier, with additions from the Human Relations movement. Organisations are machines which process information. Humans are key control elements in these machines' feedback loops, often more flexible and cheaper than actual machines (automatic control systems) but with characteristics which are less easily engineered. The science of socio-technical systems design then depends upon identifying the key sources of variance in a system, in both the machinery and the social relations, and then engineering modifications which will smooth them out. Modifications are engineered in social systems by 'installing' what these scientists see as shared values, which essentially means getting the workers to see it management's way and accept the responsibility for smooth running within the limits of what managers will allow.

The notion of preconceptualisation is still central. Managers manage the boundaries of other people's work, managers call in socio-technical designers of jobs, the designers engineer the appropriate values. But rather than being mechanists, these scientists are functionalists. They see organisations as organisms, which must therefore survive, which means that each part has a job to do. The basic problem is to define the jobs in a way which allows enough flexibility for workers to adapt to local variations, but without using their autonomy in a way which would be – the metaphor follows naturally – cancerous. The 'laws' are different, the style is different, the metaphors are different. But the basic condition of management

187

science remains in that whatever autonomies are engineered, they must leave a clear space for the masters'.

It is appropriate to note the close integration of science and power in more and more areas of life, in education, factory, school and science itself. In 1911 Frederick Taylor concluded the Introduction to his *Principles of Scientific Management* by prophetically saying that:

Fordism and the Factory

1. In-line animal dis-assembly

2. Ford takes the idea and applies it to assembly

3. The assembly of parts, and of whole cars

4. All housed in a large shed as a continuous system

the same principles can be applied with equal force to all social activities: to the management of our homes; the management of our farms; the management of the business of our tradesmen, large and small; of our churches, our philanthropic institutions, our universities, and our governmental departments.

With time,
method and
quality all measured

The design of cars
and of their
manufacture
was a whole process

. Advertising was
part of the process

8. The factory was the
    core of the
    Ford way of life

9. Ford cars and
    factories were
    exported
    worldwide

10. Working for Ford
    was also developed
    as a style of life

11. The style became
    a sign and
    an embodiment
    of the American
    way

10 **Violent Science**

DID YOU SEE the TV programmes reporting America's unmanned space programmes following the moon landings – Voyager and Pioneer? Were you held, like me, fascinated by the whole thing, much more than by the Apollo project which, with flamboyant heroicism, put real live men on the moon? Pictures of Pioneer

HEADPIECE Hardiman, an exoskeleton designed to make an ordinary GI into a super warrior. Developed by General Electric Research and Development Center, Schenectady, New York. A man is attached to the machine at feet, fore-arms and waist and his motions are amplified by the electro-hydraulic systems of the machine.

circling Venus, so real you could feel you were there, yet actually composed and animated by a computer? A wealth of insight into the Earth's geography flowing from radio-beamed robot investigations of fantastic 'weather' systems and moonscapes. The elegant trajectory of Voyager as it threaded past Venus, Jupiter, Uranus, Saturn, swinging in precise arcs using the pull of those planets to accelerate and redirect it on its programmed course. Another computer simulation, this – the real contact with Uranus is scheduled for 1986. It will happen, you can be sure. Although I trained as a scientist I'm often sceptical of 'Gee Whiz' science on the box. But the poetry and drama of this had me cold.

I didn't understand why I was so fascinated. It took a while to click . . . not the infinite wonder of nature, and not the power of machines, but the incredible versatility, panache and reach of *people*. All those workers at mission control huddled over keyboards and screens. Decisions had been made to explore the outer planets, remotely, and now the outer planets were being circumnavigated and mapped, with tremendous style. All those crewcut men logging data. The TV programmes were a celebration of our ability, collectively, to know and to respond. Yet those exciting images of exploration and the images of corporate dullness just did not seem to add up. Surely this is all harmless enough, so why should a chapter on violent science start in this way? The answer is that the ethos and the order of science is very complex indeed and there are depths and contradictions which are hard to meet eye to eye. The way I'm going to explore this topic is itself a little strange, but the topic calls for it. Rather than listing American space funding, or military implications, or talking about the machines themselves I'm going to address the question: What makes it tick? Along the way I shall point at some regrettable things – or at least, they upset me: baroque military technology, neurotic masculinity, corporatism, detached intellect coupled with brute force. And yet Voyager still fascinates. Very strange . . .

ROCKET JOCK VERSUS THE MAN IN THE CAN

By the time the first atomic bomb was tested, in America in July 1945, the war in Europe had been over for three months. In 1939 Einstein had written to President Roosevelt proposing that for deterrent reasons only the US should develop an atomic weapon. The Manhattan Project which produced that weapon had been started up under the perceived threat of enemy competition. Yet when it became clear that Hitler's Reich was defeated and that the Germans did not in fact have an equivalent weapon, research on the Bomb project was nevertheless intensified. Although the war in the Pacific was approaching an end, with a conventional invasion planned, two atomic bombs – Little Boy and Fat Man – were dropped on Hiroshima and Nagasaki on August 6 and 9, 1945. The reason was that a conventional victory over Japan would not have given such a dramatic demonstration to Russia of America's military and economic power.

The military-industrial aggression of America against the USSR continued after the war with forced conscription and massive arms spending, so that eventually Russia's Sputnik programme arose within the vicious competition of the Cold War. As Sputnik I bleeped its way through the skies on October 4, 1957, Congressmen began to ask whether Americans could sleep at night while there was the prospect of having a Red moon over their heads. They knew the answer they wanted, and the US space programme was launched. The atomic physicists may have been naive, and moon-rock collectors may – just conceivably – still be naive (as Oppenheimer said after the war, the scientists 'didn't know beans about the military situation'). But space research never existed in a power-political vacuum, any more than research on the Bomb ever did. We need to be aware just how un-innocent space research is, as a major commitment of national resources.

The USA did, in fact, already have a space flight programme and unlike the programme which eventually received massive funding it was actually a flight programme, involving test pilots and high altitude rocket-powered aircraft from the Bell X-1 up to the X-15. When the political pressure came on to get a man into space the plan which won favour was this: put him in a capsule, mount it on top of a rocket, and shoot him up there. The contrasts are dramatic. The Bell test flights were carried out within the old jockey tradition, staying up all night whoring and boozing, driving like crazy down dirt roads back to base and mounting that bucking little beast of a needle-nosed plane to fly red-eyed and bristly to the freezing fringes of space. Hands on. And to land, sometimes, by the seat of your pants. Just you against nature, and the machine was part of it too, full of unexpected tricks and wiles. The lone hero in single combat, riding Mother Nature.

This always had been a myth of course, although the aggressively masculine style was in no way unreal, for even in World War I the pilot-hero had been part of a team of service and repair workers, and the Bell projects trailed a length of tail well on the way to the present day 'teeth to tail' ratio of around one pilot to seventy ground crew. Nevertheless, like all active myths it captured some part of the reality, namely that the Air Force space programme was extremely dangerous and pilots faced that danger alone to a significant extent (as did their families). Compare this with the set-up in the NASA programme. Everything was system. The rockets, the capsules, the flight plans, were all sub- and sub-sub-divided into tiny bits, and each bit was checked and double-checked. The flights were simulated over and over so that when the day came it was hardly Big at all – just a walk out there for another run through.

The man in the can wasn't a pilot, an active flyer, at all. You went up, you came down. You got fished out of the sea, you were a goddamn hero! Projected into history. The old-dog rocket jocks couldn't understand it at first, though they soon wised up to the prestige and learnt new tricks.

A Fix For All Seasons

Not that Apollo and all the other projects weren't masculine. The technology was Army rather than Air Force, so the rockets used were the big Mercury-Atlas stages of intercontinental ballistic missiles and later the massive Saturn series; very big, very powerful, very noisy and very erect. You were under power from the start, none of this gentle mother ship stuff (the X-1 was slung under the belly of a bomber to reach thinner air, saving fuel and weight). It was a different mode of manly stuff, a kind of American thinking man's football game against nature. There was a play for every situation at the call of a number, and man-for-man marking on all – but *all* – foreseen possibilities within the repertoire of Mother Nature's trickery. A massive overkill of intellectual firepower – all those crewcuts at the computer viewscreens – rained down on any freak event that broke through the pattern. No more, the Lone Hero. This is the corporate power-in-numbers strategy, the out-ratiocination of nature through a quantitative pyramidally networked four-eyed seven-stone jigsaw-puzzling Plan. A plan to end all plans, tight, tied up and certified Double-A rational.

At least the lone-hero rocket jock was in touch, through the seat of his pants and the cold sweat on his brow, with nature. His aggressiveness was sadistic, where feelings still did count even though they were vicious, and his masculinity was assertive. There is no need to bewail its passing, but it has passed. The masculinity of the corporate man-in-the-can is different, defensive, his aggressiveness pre-emptive and cool. His drills and his armour and above all his electronically measured distance make sure that there is *no touching*. In 1925 the poet William Carlos Williams sensed this in the American grain:

> Cessation, not of effort, but of touch . . . no end save accumulation, always on the way to BIGGER opportunity . . . This makes scientists and it makes the masochist. Keep it cold and small under the cold lens . . . a passion equal to the straining of a telescope.
>
> Be careful whom you marry! Be careful for you can NEVER know. Watch, wait, study.
>
> Deanimated, that's the word; . . . Yankee inventions. Machines were not so much to save time as to save dignity that fears the animate touch. It is miraculous the energy that goes into inventions here . . . That force is fear that robs the emotions; a mechanism to increase the gap between touch and thing, *not* to

have a contact.

The violent science of corporate scientists, grey power, acts out the fantasy of untouchable, unfeeling action on a massive scale.

Rational War

Yankee ingenuity had something special to do with it, but the lesson had really been learned more recently (partly from the Limeys) in World War II. 'Science' had won the war, and 'rational war' – decision-making and weapons development through systematic research and policy-making, all under the heavy blue influence of numbers – had become a slogan of progressives who felt that a nation's ends (a democratic nation's ends, of course) should be served powerfully by the most efficient of means. In machinery (electronics, especially computers) and in metaphysics (rational management of death dealing) World War II was the birthplace of the 'electronic battlefield' which came of age in Vietnam where remote sensors dropped ever so gently by high-tech fighter-bombers would trigger bombs lovingly designed for maximal bloodiness at the drop of a footfall on a trail, all according to the California dreaming of Think-tankers right down to the last projection and rationalisation.

The electronic battlefield is now alive and well and living in the city; bugging, surveillance, national police computers, computerised 'command and control' systems linking mobile units which can deny territory to any group defined by law enforcers as an enemy: terrorists, rioters, marchers, protesters. The mentality is now powerfully institutionalised: 'We must KNOW . . . PLAN . . . COUNTER . . . everything.' Cybernetic designs aim to provide managers of business, law and order, struggles with guerillas, struggles with nature, with the requisite variety and the capability of responding to and flattening everything that the environment – everything Out There – throws up. Stability. The rational war of All-in-here against All-out-there is a neurotic war, and its ethic is material power.

The innocent space science of other worldly geography, with its techniques of ultra-remoteness is a science of a culture which worships power – power at a distance – just because it works.

NORMAL SCIENCE MEETS THE MUSCLEBOUND MILITARY

Military science is quite normal. It is normal in the statistical sense of being the norm. A government estimate suggests that one worker in every thirteen in the UK produces weapons; the engineering trade union AUEW–TASS estimates that one-quarter of its members are in defence industries; perhaps one-third of all British scientific and technical workers are defence workers. In 1981–82, fifty-three per cent of government money for research and development went to defence (total defence spending was five per cent of Gross

194

Domestic Product, twelve per cent of public spending, £12,300 m). Establishments like the Atomic Weapons Research Establishment at Aldermaston and the Government Communications Centre, Cheltenham, employ a total of 35,000 researchers and technicians. The National Science Board gives US figures for 1978–79 (*Science Indicators* 1980) as a percentage of total Government spending on research and development: energy production, economic development, health, community services and 'advancement of knowledge' between them get thirty-nine per cent; space research has twelve per cent; and defence takes forty-nine per cent. Military science is the norm.

It is normal in another sense too. In 1962 an American historian of science, Thomas Kuhn, published a book on *The Structure of Scientific Revolutions* in which he upset the applecart of orthodox Anglo-Saxon ideas of scientific progress. Challenging assertions and critical experiments – revolutionary changes in ideas – were, Kuhn claimed, occasional phenomena of science; most scientists all of the time, and all scientists most of the time, are engaged in a kind of jigsaw-puzzling activity within an accepted framework which Kuhn called 'normal' science. Whatever its value as a general way of seeing science in history this image of science as day-by-day filling-in of a rather complicated crossword puzzle appealed strongly to many working scientists, reflecting as it did their own experience of how scientific work is done today. To name the crossword grids and the techniques used in solving them, which scientists are trained into and accept as the form of the status quo, Kuhn gave the word 'paradigm' a special sense (twenty-six senses, according to one critic). Scientists inherit a paradigm, work within its prescribed intellectual models and procedures, use its implicitly specified tools and techniques, and add on to the 'body of knowledge' which accumulates as a product of this organised labour. 'Accumulate! Accumulate!' is Moses and the Prophets for 'normal' science; progess is MORE – more results, more citations, more students. Although Kuhn himself suggested that a social-psychological process governed entrenchment and change in science, it seems obvious that in its 'normality' present-day scientific research is determined by economic-political alignments. Military research is the central instance of this.

Baroque Technology: The Paradigm of the Egg-Laying Sow
In military science Kuhn's 'paradigm debate' becomes debate over weapons systems. One group of researchers, working for a particular group of firms which manufactures one system, 'debates' (that is puts forward highly complicated quantitative arguments) against other groups equally committed to other systems. Positions are deeply entrenched; soldiers seek to justify past and existing military roles and policies, paymasters seek to retain levels of military estab-

lishments, corporate heads seek to sustain manufacturing capacity and profit levels, researchers seek continuance of their personal contracts and lines of work. There is intense competition between firms and between arms of the forces, which leads to minimally radical innovation by merely tinkering with existing forms of technology and strategy. The results are tremendously sophisticated and elaborate weapons systems, feats of immense ingenuity, talent and organisation which can inflict unimaginable destruction. A single airborne missile sinks a ship. A neutron bomb is a dirty weapon carefully designed for maximum human deaths and minimum destruction of hardware. 'Progress' in this world is more staff, more money, more controls, more facilities.

As expense rises there develops a pressure towards multi-purpose systems and multi-national collaboration. With the F-111 airplane this was overdone, and the US Navy had to cancel because the plane just wouldn't do everything and still do any of it well. The Anglo-German-Italian MRCA (Multi-Role Combat Aircraft) has been likened to a sow which lays eggs, grows wool and gives milk. Britain wanted a long-range strike and anti-bomber aircraft; Germany wanted close air support; Italy wanted air superiority against fighters. What they got was something like a very expensive low-level bomber for nuclear strike. Financial and profit-making considerations thus reinforce the tendency towards more of the same. Military research becomes concentrated and internationalised because it is so heavily capitalised, and this enhances and entrenches the tendency of 'normal' science to get hold of an idea which can be broken into a thousand small pieces and milk it for all the control it can yield.

The technology itself becomes decadent, incapable of limited use, extravagantly costly, difficult to handle, demanding of maintenance. Military airplanes are less reliable than commercial planes. Tanks break down more often than tractors. Warships are out of commission more often than merchant ships while costs of maintenance, spares, repairs and fuel increase. Militarily, many of the sophisticated facilities are redundant: once you're in the 'envelope' of an enemy plane's missiles you're dead, no matter how fast or manoeuvrable your superfighter. This is the 'baroque arsenal'.

Musclebound

Similarly, the institutional sclerosis of the 'military-industrial complex' can be seen as a sort of cancerous normal science, become massive and visible by the growth of financing and staffing committed to endless 'trend innovation', change without development. With more workers required to produce the results and the weapons, more workers to operate them and more components to assemble (supplied by more small sub-contractors), more specialisation develops in the division of labour and more control (through more controllers) is applied. Variety of systems declines, as does the

196

variety of military tactics, leading to more integration of strategy. By merger or collaboration, the number of prime contractors declines while interdependence through interlocking contracts goes up, resulting in further concentration of the power of capital. All these tendencies lock together to reinforce the hierarchy, as well as consolidating a multi-national convergence of military-industrial organisation. The outcome is a very rigid technological and economic system. So large that unmanageability is an ever-present nightmare confronting research managers and military commanders, the organisations spawn intra-organisational game-playing and incestuous career patterns. By 1960 there were six hundred and ninety-one retired US generals, admirals, naval captains and colonels on the payrolls of the ten largest military suppliers; 'advisers' flow the other way.

This sounds like Kuhn's description of a 'paradigm' in crisis. Military science *is* a crisis but it is not primarily a perceptual crisis, as Kuhn proposes in his model of 'paradigm change'. It is a political-economic crisis of social control in that a segment of social practice is out of control, internally and externally. The crisis assumes very dramatic and visible proportions in military research because of the financial muscle possessed – that is, appropriated – by the institutions. But it can be seen as merely the logical extreme of normal science, in which progress becomes progressive disease. 'Paradigms' acquire muscle: personnel, laboratories, equipment, connections within the State administration – in non-military as well as military science. Paradigms acquire muscle through manufacturing capacity and market shares; this applies to most research too, about seventy per cent of scientists and engineers in the US work in commercial laboratories. It hardly seems an accident that at the same time as the American academic, Thomas Kuhn, was projecting his up to the minute model of 'normal' science back into the history of science, like a good bourgeois intellectual, the American president and old-fashioned general, Eisenhower, was becoming edgy about the entrenchment of interests in research.

In 1961, the year before Kuhn's book was published, President Eisenhower warned, 'In holding scientific research in respect, as we should, we must also be alert to the equal and opposite danger that public policy could itself become the captive of a scientific-technological elite.' It was then that he coined the term, 'military-industrial complex'. In a sense the academic had more insight into contemporary science than the president, working, as he did, in California where two-fifths of jobs depended directly or indirectly on military contracts. He could see that the tendency towards sclerosis was 'normal'. If large concentrations of capital and entrenched military interests are added to the 'puzzle solving' fetish a baroque edifice of power results which – and Eisenhower was wrong here – is not the opposite of normal scientific research, but its apotheosis. Multinational capital and militarist teeth-gritting determination just supplied the Charles Atlas and the muscle to normal science's seven-stone ego and its fantasies of control.

IN CONTROL – OUT OF TOUCH: THE CORPORATE IDENTITY OF SCIENTISTS

Power. Scientists and engineers do not wield power in the military-industrial complex as individuals or even as a distinct lobby. Managers may or may not often kick sand in their face, but seven-stone science is up against the big boys. The technicians are, however, central in the articulation of political power and the worship of power – effective control from a distance – makes them natural participants in manipulative politics. This 'style' is linked to a willingness among career scientists to accept a definition of personal identity in terms of career competition and intellectual rigour, within which caring relationships (that is, conventionally, home life) are entirely subordinate, if not actually suppressed.

On one hand, scientists in active and prestigious areas of research such as space research, find a source of great personal satisfaction in their work. They enjoy it when they get the right results, they get status and respect from their colleagues. But on the other hand there is the constant anxiety about being scooped, or failing to get the right results (and the promotion, and the new mortgage), or falling behind and flunking the continual regrading imposed by 'peer review' and missing the plum contracts. Once they are let out of the swaddling bands, at the age of around twenty-six, they are supposed to spread their wings and fly to a safe spot and settle in and Do Well by their mid-thirties. This is a description of the life of academic scientists, but there is little to distinguish it from the rat-race existence of the Organisation Man. Like him, too, it seems that workers in 'hot' fields of research devote more time, energy and emotion to their work and their colleagues than they do to their families and the rest of the world 'outside'. Saturn can be more real than their own neighbourhood swimming pool.

Male researchers generally have an imperfect consciousness of this. One moonrock physicist was able to say blandly, 'I talk to my wife all the time. I think she has a minimal understanding of my work. Science is my total existence, 365 days of the year, no vacations.' Many clearly regard their wives as lacking understanding, so that, 'My wife makes a face whenever I go to the lab at night or on Saturday, but she "understands".' Some may accept that their commitment to research imposes conditions of existence on their spouses which are disagreeable, 'If colleagues come to dinner and talk about science, my wife feels left out. She's put off by the feeling of intellectual superiority that most scientists have, especially physicists, who she feels are affected, arrogant and cocky.' 'She' happens to be a professional counsellor with many scientist clients. Whatever the awareness, male moonrock researchers call upon their wives continually for emotional and practical servicing: 'My wife has no real background in science but has a very strong aptitude for understanding people and situations, which makes her an extreme asset to me. I usually talk to her about my work every evening, about daily frustrations, and occasionally, daily accomplishments. My wife might prefer me not to be a scientist; she doesn't like me spending so much time away from the family.' Normal science is corporate science; the scientist's wife is a corporate wife (an 'asset'); and normal life is subordinate to the demands of the institution.

The 'professional' cut off of scientists keeps them from accepting some of the facts of their emotional relationships with what convention calls 'loved ones'. According to some research results which have now become a cliché (originally published by the psychologist Liam Hudson in *Contrary Imaginations*), this professional coolness has roots which are found much earlier in the 'converging' and 'diverging' tendencies of thinking in schoolboys. Most Arts specialists are divergers, most physical scientists convergers. Reactions of convergers to controversial issues tend to be stereotyped and compartmentalised, which may, by ignoring conflicts, serve as a defence against anxiety. Convergers tend to be authoritarian, and to keep their feelings to themselves. 'With the single exception of extreme or "morbid" violence, every differentiation reveals the converger as the more discreet in the expression of what he feels . . . Many convergers, one suspects, do not stifle strong, disorganised feelings – they fail to experience them.' It is certainly the case that contrary imaginations, centering in the experience of feelings, show themselves in the adult lives of physical scientists. The following are quotations from American East-coast high energy physicists and their wives:

Spouses quoted in Ian Mitroff, Theodore Jacob & Eiken Trauth Moore, 'On the shoulders of the Spouses of Scientists' *Social Studies of Science*, 1977, 305–27.

Deep tensions such as these are very difficult to negotiate, they call for time, and touch, and real emotional work. But the world of work in the corporation is much more 'objective'. Becoming functionally identified within the organisational 'machine' makes life much more easily liveable – if less lifelike. When you abandon your self to its purposes and definitions, the Organisation responds like an extension of your (reduced, operationally defined) self: it fits like seven-league boots.

The emotional strategy of convergent thinkers, who become 'hard' scientists, maps the strategy of knowledge in reductionist science; formal, narrowed language with a purposefully reduced range of expressiveness, conventional definitions of roles and existence within the uncertain real world. In turn, this maps the emotional strategy of corporate life. Keep the real world at arm's length, look at it 'with a passion equal to the straining of a telescope', decouple yourself, submerge your perception in the definitions of the organisation. A reduced career identity in reductionist science is the counterpoint of corporate existence in society.

DISTANCE, FORCE, HARDNESS

To return, via reductionism, to the scientists; the space scientists and their wonderful robots, exploring the universe with fantastic reach, innocently seeking strength through toys. Robots are the ideal of hard science because they make processes entirely 'objective', that is, only objects are involved. With robots on the job, from transducers to automatic chromatographs to space-exploring automatic laboratories, there are no people taking part, measuring and sampling and calculating. These are the standard to which the human processes of planning, selecting and evaluating can only aspire, until the project of 'artificial intelligence' pays off with a robot scientist. 'Objective' robotised processes are the ideal of science because they show unequivocally that man is IN CONTROL. When the Viking lander flew to Mars and sat there on the red sand photographing and scooping up bits of the landscape it was a demonstration not just of the adequacy of the theories that scientists had about the composition of Mars soil, but fundamentally of the power of remote handling technique.

To see this embodied at incredible expense in the Viking and Voyager and Pioneer programmes is only to see the mode of reductionist science writ large. Success in modern science has been prominently claimed by those research programmes which sought control of the invisible – of molecules and atoms and parts of atoms, forcing them to show themselves, to leave a spoor, to trip a wire. Science has succeeded by leaving the world of ordinary life, where meanings and feelings are inescapably part of what we see wherever we look, and migrating to a constructed world in which a grey language of black and white distinctions could be made to do. In everyday life

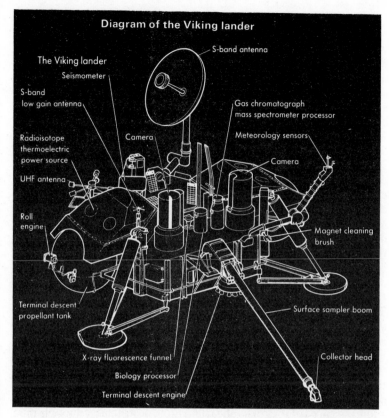

Diagram of the Viking lander

S-band antenna

The Viking lander

Seismometer

S-band
low gain antenna

Gas chromatograph
mass spectrometer processor

Radioisotope
thermoelectric
power source

Camera

Meteorology sensors

Camera

UHF antenna

Roll
engine

Magnet cleaning
brush

Terminal descent
propellant tank

Surface sampler boom

X-ray fluorescence funnel

Collector head

Biology processor

Terminal descent engine

we can 'know' about someone or some situation because we read a
signal or sense a mood or anticipate a response. We can communicate it in other ways and other languages, including body language
and the language of silences. But we could not express any of this
in the reduced language of science. Reductionist science makes this
kind of knowledge impossible for itself by creating a world of
non-natural objects, taken outside their normal contexts or actually
created in the laboratory. The knowledge is powerful in the lab, and
in similarly artificial places of reduced purpose and technical control,
which says nothing – and scientists say nothing – about the value
of this way of knowing in relation to the other ways.

In the constructed, distanced, hard-edged world of reductionist
science we cannot know until The Test. The test is in fact a contest,
between Intellect and Nature. And when the result of the manipulation of nature is what it was predicted to be in local operational
terms, this shows that we are IN CONTROL, in local operational terms.
The greatest degree of control in science has been demonstrated in
a world where touch is impossible, which is to say that all action is
force, action at a distance. The project of reductionist science might

then be put this way: the triumph of abstract intellect depends on creating a world in which physical force, fiat, is the only mode of acting.

What's Wrong with Reductionism

There are two things wrong with that. The first is to do with culture. Nature is not a person and has neither meaning nor feelings in itself. But that is beside the point because nature 'in itself' is never part of our experience; we are bound to experience nature as part of human experience. This means that it always carries meanings, responds to purposes, and this in turn means that when studying and acting on, and reacting to nature we need to be conscious that we are negotiating meanings and material relationships with ourselves and other humans through culture. We act *in* nature, as parts of an apparatus, and ecology as a scientific approach explores our relationships in this respect, while necessarily challenging reductionist methods in doing so. But we also, even when doing 'hard' ecological research, act in a socially constructed system of cultural relationships invested in 'things'. Reductionist sciences' commitment to searching for 'things' where there seem to be none is itself a case of this; reductionism is a cultural project, and the things it creates to study are cultural objects.

The second dangerous aspect of the intellect/force dualism is that it exists in, and is a basis of, a historical system of elite power. The ex-scientific adviser to British governments, Lord Zuckerman, insists in *Nuclear Illusion and Reality* that 'nuclear weapons are not weapons of war', and he does this through stressing the material consequences of their use, the deaths, the destruction, in opposition to the distanced, hard, reductionist language of megatons and megadeaths. But if they are not weapons of war, what are they? For one thing, they are materialisations of State power, the expression – as massive Force – of the identity of the cultural elites which govern and rule. They are also profitable commodities, which feed the capital of firms which deal in arms, computers, construction. Thirdly, they are the apotheosis in massive force of the massed intellect of reductionist, corporate research workers, and as such they represent a powerful force of male neurotic anxiety, the relentless convergence of inquiry in a search for the power of control over everything out there.

The coupling of a brittle, emotionally reductionist identity – the identity of the 'converger' – with a military-industrial calculus of terror and profit is as 'natural' as the couplet of intellect and force in reductionist method. Like all of nature, it is a historically specific cultural phenomenon. But also, like nature, it is deeply refractory, it hangs together for a powerful complex of reasons which, while not 'natural', are so near to it that changing them is a massive challenge. That kind of conscious and deliberate massing of effort

202

today tends to underlie neurotic projects rooted in anxious males' emotional cut-offs and fantasies of remote sensing and control, searches for the total artificial Life Support System, a man-made womb which will preserve physical existence at whatever reduced level in the face of every possible violent response of a violated nature. The systems flow from the drawing boards of military-industrial designers, closed, armoured, hardened, wired for sound, electronically pulsing. The tank. The fallout shelter. The hardened command post, silo, satellite, warhead. The space capsule. The saturation diver's pressure vessel. The atomic sub deep beneath the polar ice sheet. The aim is entirely unnatural life in strenuously

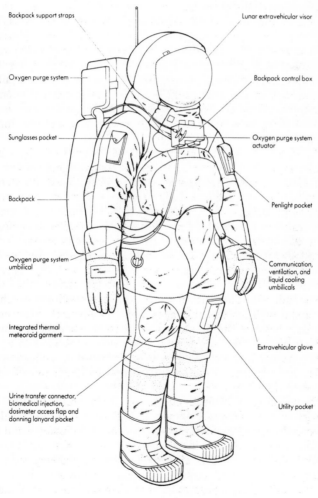

Diagram of Apollo lunar spacesuit.

sought extremes at and beyond the frontiers of human existence. These are not the means of life, in any general or generous way of living. They are reductionisms' reduction of life – through the power of capital and of the States' ruling elites – to tableaux of prediction and control.

DOES SCIENCE NEED TO BE LIKE THIS? The question is not actually to the point. Of course it is possible to imagine something different. The practical question, because it directs attention to the culpability of us as well as them, is: Do we need science to be like this? Answers are contradictory as are relations between working people and experts, use values and capital. The answers are very complicated as are our needs and our fears.

It would be absurd to try to answer the question here. But it can be placed in a concrete context. We may be innocently fascinated by the control that people have over machines – we may be Voyager watchers. Do we also, unselfconsciously, have a fetish of control over people's behaviour? In 1979 members of the American public were asked which of the following fields of scientific research they were most willing to see restricted: engineering new life forms, discovering life in space, extending the human life span, controlling the weather, and detecting criminal tendencies in young children. They were more supportive of those projects in the order as listed; most people, more than eight out of ten, said that scientists should study ways of detecting criminal tendencies in young children. In terms of needs and fears, there is powerful support for that project of scientific research. But 'hard' scientific research is a project of prediction and control. What controls will the public sanction? What controls will flow 'naturally' into practice from research? What neurotic needs will be entrenched behind the research front and the technical fixes?

Just chew on this incomplete list of means of control which have been or soon could be used: sterilisation of potential parents, compulsory abortion, 'prison' schools, drug 'therapy' for 'hyperactive' schoolkids, psychosurgery (taking out centres of insurgency in the brain), implantation of electrodes in the brain coupled with a computer surveillance system which warns 'guardians' of impending bad acts so that the 'patient' can be electric-shocked into being good; and tomorrow, microcomputers grown from living matter, which can be surgically implanted in deviants' brains to provide day and night surveillance and control. At what point will our fear stop, in its violation of others' integrity? At what point does scientists' anxiety start to make them violate Nature's integrity? How do we make a culture in which our needs no longer require science to be like this?

204

Swords to Ploughshares

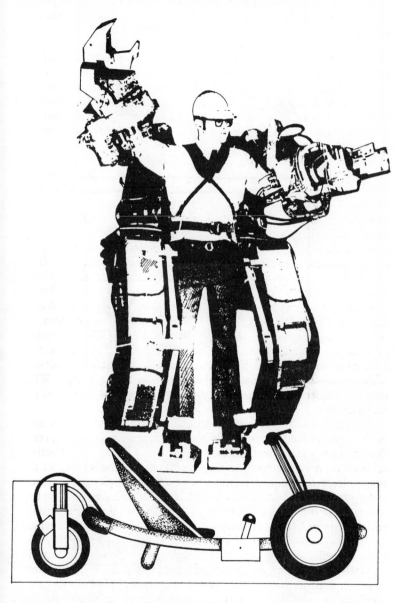

The hob cart is a machine designed and made by Lucas Aerospace shop stewards to give mobility to physically handicapped children. The shop stewards of the Combine Committee proposed that such products were a possible and desirable alternative to the sophisticated military machines produced by the arms industry.

LUCAS AEROSPACE WORKERS

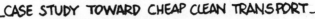

CASE STUDY TOWARD CHEAP CLEAN TRANSPORT

1. ORDINARY CAR

HIGH FUEL INPUT HIGH POLLUTION

FUEL IN

DRIVE WHEELS

GEAR BOX

ENGINE

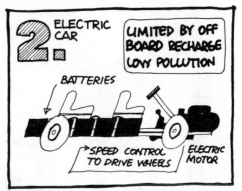

2. ELECTRIC CAR

LIMITED BY OFF BOARD RECHARGE LOW POLLUTION

BATTERIES

SPEED CONTROL TO DRIVE WHEELS

ELECTRIC MOTOR

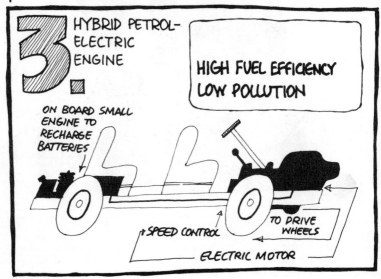

3. HYBRID PETROL-ELECTRIC ENGINE

HIGH FUEL EFFICIENCY LOW POLLUTION

ON BOARD SMALL ENGINE TO RECHARGE BATTERIES

SPEED CONTROL

TO DRIVE WHEELS

ELECTRIC MOTOR

LUCAS AEROSPACE WORKERS

Workers Discover the Meaning of 'Multinational'

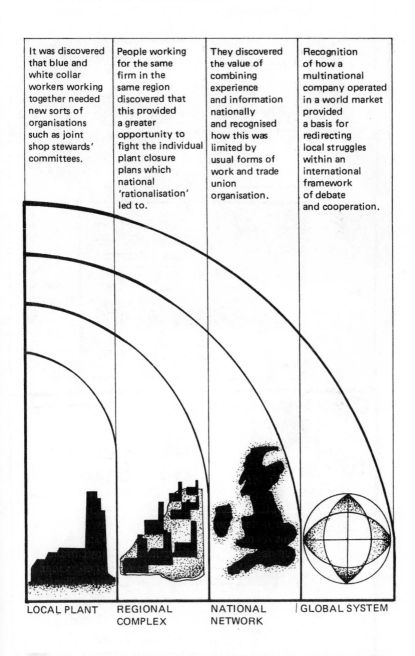

| It was discovered that blue and white collar workers working together needed new sorts of organisations such as joint shop stewards' committees. | People working for the same firm in the same region discovered that this provided a greater opportunity to fight the individual plant closure plans which national 'rationalisation' led to. | They discovered the value of combining experience and information nationally and recognised how this was limited by usual forms of work and trade union organisation. | Recognition of how a multinational company operated in a world market provided a basis for redirecting local struggles within an international framework of debate and cooperation. |

LOCAL PLANT   REGIONAL COMPLEX   NATIONAL NETWORK   GLOBAL SYSTEM

EIGHTY PEOPLE DIED in the 1889 Armagh train disaster, one of an appalling series of railway accidents caused by runaway trains. Yet the technical means for preventing this kind of accident had been in existence since the 1860s – the continuous vacuum brake. If a coupling snaps and a section of the train breaks away, the loss of vacuum applies the brakes automatically. So why was the system not in use? The simple reason – and the one given by the owners – was competition. Each company argued that it would put itself at a competitive disadvantage if it spent money to install the system and other companies did not follow. Consequently no one did and it took a whole series of accidents to force the issue morally, although an Act of Parliament which was passed a full ten years after the Armagh disaster covered only brake systems on passenger trains. Commercial disregard for safety was even worse than at first seems, because a Railway Inspectorate which had been in existence since 1842 was quite unable to bring a humane logic to bear on the railway owners until the deaths of consumers forced a public outcry. Must the consumer always beware? Is this, fundamentally, the relationship between consumers, products and technical knowledge? Can experts modify or change it – why not? These are this chapter's questions.

It might seem that the situation has changed radically over the past hundred years. Consumer movements, or movements in the avowed interests of consumers, have had some quite powerful and visible effects, especially in the past decade or so. Ralph Nader, America's best-known consumer investigator, led 'Nader's Raiders', spearhead of a protest movement against the polluting effects of the automobile, with all the hard sell and ballyhoo of American marketing and promotion. 'Unsafe at any speed' was their line on the Ford Pinto which was subsequently taken off the market. Today some of the most careful design and sophisticated technology in the auto industry is directed towards satisfying consumer needs as highlighted by campaigns of this kind: crumple zones for protection in crashes, microchip circuitry for ignition and emission control. In Britain one success story is CAMRA (the Campaign for Real Ale) which has materially changed the map of beer drinking, not only so that you can see the difference in the beers behind the bar in most pubs, but also in the distribution of brewing capacity and market shares between the big nationals and smaller local concerns.

Or think of bread. Marketing's sensitivity to consumers' needs, or at least, wants, has transformed the array of fresh-baked bread in big supermarkets or the high-street hot-bread shops. How different to the situation in mid-nineteenth-century London, as revealed in the House of Commons Blue Books on the adulteration of articles of food (1855–56) and Dr Hassall's *Adulterations Detected*. Marx wrote:

Englishmen, with their good command of the Bible, knew well

enough that man, unless by elective grace a capitalist, or a land-lord, or the holder of a sinecure, is destined to eat his bread in the sweat of his brow, but they did not know that he had to eat daily in his bread a certain quantity of human perspiration mixed with the discharge of abscesses, cobwebs, dead cockroaches and putrid German yeast, not to mention alum, sand and other agree-able mineral ingredients.

The House of Commons Committee found that free trade in food essentially meant trade in adulterated goods, or 'sophisticated' goods as they were ingeniously called. Three-quarters of London bakers were 'undersellers' who traded in bread adulterated with alum, soap, pearl-ashes, chalk, Derbyshire stone dust, etc., etc. The working people who bought the bread knew how bad it was, and how bad it was for them, but being paid *after* their week's work, were bound to accept whatever the local baker or chandler's shop chose to supply on credit. As a witness to the House Committee underlined, 'bread composed of those mixtures, is made expressly for sale in this manner'.

Compare this with modern sophisticated plant-baking techniques. Most plant-baked bread uses the Chorleywood bread process which essentially replaces three hours of fermentation in the traditional baking process by a few minutes of intensive mixing. A brew of water, yeast, sugar, salt, skim-milk powder, soya flour, calcium carbonate (chalk), oxidising agents and 'yeast nutrients' is fermented before being added to the flour, together with some fat. A very high-speed mixer smashes it together in a couple of minutes. Not only faster and therefore cheaper to produce, but also easier to oversee and therefore even cheaper to produce, because skilled bakers are no longer needed to make judgements about the dough which is fluffier and therefore still cheaper to produce, because the same amount of flour goes further. Chorleywood bread is three-quarters *air*.

The ploy is similar to the one used by sophisticated manufacturers of poultry, bacon and cooked meats. They add sodium tripolyphosphate which makes the flesh absorb more water, and sometimes bounce it around in a tank of brine just for good measure. Have you noticed the little bubbles in the pressed ham shoulder that you buy? Result: you buy added water by the pound, at the same price as meat. In their own ways, modern foods are just as 'sophisticated' as the undersellers' bread, and for the same reasons of profitability. But we eat bread and meat adulterated with air, and with water.

The Role of Experts

The role of technical knowledge, and of experts, in all this needs careful consideration. It is worth noting that over a century or so the experts changed sides. Dr Hassall in the 1850s 'discovered' the adulteration of bread and campaigned against it, while the adulteration itself needed no supporting expertise because bakers could just tip in anything which might escape a casual glance. In the 1950s it was experts – food technologists – who produced the new techniques responsible for Chorleywood bread's air-adulteration. A second thing to note is that even though it took an outsider – Dr Hassall – to discover the adulteration of bread officially, it had already been 'discovered' by the working people who had to eat it. (Similarly, army doctors were not the first to discover the terrible malnutrition and physical disability of recruits in World War I). It was outsiders, concerned members of the Victorian middle class who took up bakery-workers' complaints that they were cruelly exploited through overwork, and pressed the enactment of legislation for shorter working hours. Marx records this but does not mention what effect there was on the price of the workers' bread. Did the improvement of workers' conditions and consumers' nutritional standards take cheap bread off the market because bakers couldn't make a profit? This, presumably, is another 'discovery' that working people would have to make before experts scientifically sanctified it and began yet another parliamentary lobby.

In general, even when they are supporting workers, the involvement of experts is problematic. There is a case of a group of workers who handled pesticides and who realised that numbers of them had become sterile. They fought for recognition of this fact, eventually got an enquiry set up, and won it. But as far as the literature is concerned it is the experts testifying to the enquiry who 'discovered' the 'side effect' of sterility. Studies of working people and work, whatever else results from them by way of social action, almost certainly end up as career capital for the expert, in the form of publications to be cited when applying for jobs, tenure or promotion. Experts and professionals have their own interests and their own commitments arising from their occupations, and this can mean that when they are involved in a campaign they appropriate it for their own special interests.

What, then, are working people to do when they find themselves having to resist violations of their life or health through products, or when they have to refuse 'progress' in new technologies, techniques and working conditions because it damages interests that are entrenched in their existence as wage-workers? Can they appeal to experts for help when experts are manifestly part of the problem? Consider some of the many issues: opposing nuclear weapons proliferation or nuclear waste disposal, new airports or motorways which destroy the quality of an environment, new machinery and processes in the workplace which destroy jobs or endanger health, racist claims about intelligence, sexist attacks on women's suitability for 'men's jobs'. In campaigns like these, experts are always on the other side, if also sometimes on ours. What sense does it make for consumers – outsiders – to appeal to science against science? Is there any way of telling which 'science' is on your side?

The simple answer is NO. There is no clear-cut way of telling the difference between 'good' science and 'bad' science, whoever you are and whatever your interests. What is possible, however, is to develop some understanding of the grounds on which different assessments make sense, when experts get involved in trying to make science work for outsiders. The way to tackle this is through unpacking some proposals that have been made by scientists for 'consumer research'.

CONSUMER RESEARCH

In the period just after World War II when a Labour government was in power in Britain and scientists felt emboldened by the contributions they had made to the war effort, progressive scientists in the Association of Scientific Workers (the AScW, a trade union) began to develop a policy for 'science and the nation'. They published a book of that title in 1947, and in it, sandwiched between chapters on individual industries and chapters on scientific research in universities, is a small chapter entitled 'Consumer Research: How to Assess our Requirements'. The main proposal was that a Consumer Research Council be set up on similar lines to the existing Medical Research Council, and that under the Council there should be a liaison, research and publicity organisation which would discover 'what the people need' and make these needs known as a guide to an integrated national production programme. The AScW felt that 'The assessment of the needs of consumers and the protection of consumers from fraud must become one important section of the research of our country,' and to this effect had pressed a resolution at the 1945 Trades Union Congress outlining the measures required. The general tone of the commitment was expressed by one AScW activist in contrasting consumer research with market research, 'Market research is definitely tied up with a producer. It is not the function of market research to protest against

MANAGERIALISM

AN AMERICAN EX-TROTSKYIST, James Burnham, published a book in 1940 called *The Managerial Revolution*. In this he asserted that capitalism was being destroyed, because of its inefficiency in managing the expanded forces of production which constituted modern societies, but that what was replacing it was neither socialism nor any form of democracy. Burnham saw a new kind of centralised, planned society. Dismissing his arguments, the English socialist author George Orwell cast some interesting light on the kinds of people who took Burnham seriously:

> Burnham's theory is only a variant – an American variant and interesting in its comprehensiveness – of the power worship now so prevalent among intellectuals. A more normal variant, at any rate in England, is Communism. If one examines the people who, having some idea of what the Russian regime is like, are strongly russophile, one finds that, on the whole, they belong to the 'managerial' class of which Burnham writes. That is, they are not managers in the narrow sense, but scientists, technicians, teachers, journalists,

broadcasters, bureaucrats, professional politicians: in general, middling people who feel themselves cramped by a system that is still partly aristocratic, and are hungry for more power and more prestige.... That a man of Burnham's gifts should have been able for a while to think of Nazism as something rather admirable, something that could and probably would build up a workable and durable social order, shows what damage is done to the sense of reality by the cultivation of what is now called 'realism'.

Conditions have changed in some respects since Orwell wrote this in 1946. Many of these people now do have more prestige, and some have more power, though they are still 'middling'. Aristocratic values have come to some kind of accommodation with the values of meritocracy. But long after the 'Technocratic Era', the technocratic ethos is still active. 'Realism' – the technical fix, the submission to 'fact' – is still a powerful cultural-political force. What would Orwell find to say differently today?

---

the sale of trashy goods.' The radical scientists of the AScW wanted to see a partisan research organisation, employing research scientists, who would stick up for the needs of consumers.

The proposed research functions fell into two categories, which we can call science *to protect* the people (or 'watchdog science') and science *for* the people.

Watchdog Science

In this connection the State is seen as protector, offering some kind of impartial analytical service or policing function in relation to goods on sale, and within the terms of legal obligations. The 1945 TUC resolution described the function as 'to promote factual investigations into the utility, efficiency and cost of advertised consumption goods such as domestic equipment, food and medicaments'. The AScW suggested an Inspectorate, by analogy with the existing Factories Acts' organisation, which would check performance against specification. This was an extension of a principle going

213

back to controls on food and drugs (vested in local government and Public Analysts) and further, to legislation of the previous century dealing with specific occupations (lighthouse keepers, canal boat workers) and public health.

The central idea is, in fact, the same notion of 'service' which was the watchword of progressive middle-class liberals active in giving form and purpose to the State apparatus, the Civil Service. The radicals of the 1940s were at pains to specify that consumer researchers should not offer or be asked to make decisions or take responsibility for them. They had to provide information and guarantee its accuracy, but dissociate themselves from its implications in the context of policy and action – the very essence of liberal neutrality, and a position derived from the ideology of 'neutrality' in science.

The limitations of this kind of institution are fairly obvious. The Railway Inspectorate had no teeth in the 1880s. Even since the Factories Inspectors were given some by the Health and Safety at Work Act of 1974, workers have found it sometimes less than easy to get the watchdog to growl, let alone bite. One of the problems here is professionals' ideas of neutrality. The 'neutrality' image of scientific opinion is a double bluff when deployed by those with power, as in the case of Lord Todd who 'neutrally' discussed science education (chapter 7) or controls on genetics research (chapter 6). In that connection it means 'I'm only advising so don't pick on me – but you'd better do what I say because I'm a scientist and it must be correct.' This line requires real institutional weight to back it up, and Lord Todd could marshal that. Those without real power, however, find themselves in an absolute quandary when they try to interpret their roles in terms of 'neutrality'.

Thus there exists, for example, what OR scientists call the OR Dilemma. Surveying more than twenty years of OR in a very large group in the State sector, a manager sadly wrote: 'Thus we come again to *the* OR dilemma, to be involved with management in the actual solution of its problems, and at the same time to be able to withdraw from the fray and dispassionately study what is being done so that major improvements can be made in the future.' In the cut and thrust of politics, 'detachment' is the kiss of death, partisanship the only way of actively entering the action. To recommend 'recommendation' as the mode of intervention by scientists into politics is to back a loser – a loser with integrity, but a loser.

The AScW did envisage, in fact, that the consumer protection agency would have the power to prohibit the sale of harmful goods as well as wasteful, useless products. Even so, vacillations of Ministers of Health and their Committee on the Safety of Medicines, for example, show that there are active insider interests inside the 'watchdog' apparatus, and although Health and Safety inspectors can shut down a factory they are also obliged in law to take into

214

account what the owners claim are 'reasonable' excuses for letting hazards remain. The question of law is at the heart of this issue. First, law-making generally bends over a long way towards property owners and their entrenched rights in opposition to non-property owners, and most consumers, as working people, are not property owners in any significant sense. And second, before a watchdog institution can be established, the damage done by some other already established institution has to be proven; not just suffered, but officially discovered, documented, and put through various parliamentary procedures, formal and informal – lobbying, and all the rest of it. There are too many professionals along the way – professional researchers obviously, but more importantly, professional fixers and log-rollers.

The 1940s radicals noted that in the USA watchdog functions were already being carried out by private non-profit corporations which fed research reports to subscribers. State funding rather than private funding was specifically part of the AScW brief, so that open publication of findings could be assured. Of course, it hasn't happened. What exists in Britain is the *Which?* organisation which is . . . a private non-profit company feeding research reports to its subscribers. The subscribers are thus doubly consumers – of products generally, within the range of middle-class income, and of *Which?* services too. It is clear that an institution like an ombudsman or an assay office, whose function is to investigate grievances or questionable objects which people present for assessment by paid investigators, is a necessary one at many levels in a democratic society. But its limitations as a State institution are also clear in that it requires submission to the law and – more materially – the processes and personnel of the law, from Members of Parliament, through civil service administrators and technical staff, to the judiciary. 'Watchdogs' are fed and kept in kennels on the land of 'experts', and if working people want protection on their own ground they must invent other institutions for the purpose.

SCIENCE FOR THE PEOPLE
More active than the watchdog role was the second proposal of the 1940s radicals. The TUC resolution was, 'To determine, through surveys and other means, the quantity, type and standard of goods required by the consuming public. . . [The Consumer Research Council] should be required to publish its findings, which should be regarded as privileged comment within the laws of libel.' In this connection the scientists felt that they were pressing for a new State function rather than expanding an established one. They wanted some means 'for ascertaining the consumers' real needs, for transmitting these to manufacturers, and for ensuring that the finished articles are reasonably fit for the functions they are to perform.' That is, they wanted to create an apparatus of feedback between

215

researchers and manufacturers of goods in a 'mixed' economy, through which researchers could project consumers' real needs, as scientifically 'discovered'. Reading between the lines it seems that 'consumers' was another way of saying 'working people', and that those who proposed the Consumer Research Council were authentically committed to advancing the interests of workers rather than employers. Yet the vagueness left too much ambivalence.

Note, for example, the softness of the partisanship of the director of one of the Cooperative Research Associations. These were units funded jointly by the State and private industry to develop or modify manufacturing techniques. In 1948, giving one of a series of public lectures for Manchester businessmen on the usefulness of OR, this director asked rhetorically:

> Whom do we serve? . . . The only solid basis on which to erect our research structure is the basis of service to the community . . . We pursue all those enquiries which are calculated to improve the service of the industry to the community, and then you can get on with your jolly old private enterprise as hard as you like, and compete as hard as you like in the transference of that into public service.

Seen this way, consumer research was done *for* the people, and measurements were done *on* the people, literally in this case, because the products in question were boots and shoes. The functional nature of this kind of feedback link in a mixed economy was well perceived by the chairman of the series – no partisan of working people – who smoothly observed during his summing up that the research director's doctrine, 'which was morally right and socially right, was also right from the mundane point of view of mere profit-making. I like the combination of idealism and practical profit-making.' A powerful and well-tried mixture.

'Consumer research', as proposed in the 1940s, was in effect a form of *producers'* research. Many of the consumers whose needs were addressed were themselves capitalist firms. A process of research into products would have assisted the more intimate appropriation of people's needs by capital, adding to the knowledge stock of firms at the state's expense – a hidden subsidy. In the absence of clarity about radical differences in needs and interests between 'consumers' and 'producers', consumer research amounted almost to charity, the good old-fashioned soup-kitchen. It is a hard criticism to make, because socialists in the 1940s really thought that power through the State and the Labour Party would change things radically, but the proponents of consumer research just did not understand hegemony. Forty years on it is hard to know whether we have a better practical understanding. A couple of examples may help us to tell.

Liberated Time in an Italian Multinational
A group of scientific and technical workers at the Castellanza branch
of the Italian multinational chemicals corporation, Montedison,
were involved in producing the radical science publication *Sapere*.
They used their experience and their location as a research group
within Montedison to pursue some enquiries, as part of a general
commitment to science for the people. That is, they liberated some
of their time as Montedison wage workers in order to help other
workers in and outside the company, using the know-how and the
facilities that they were paid to use. They chose to use their own
work in an alternative way. In a non-State 'watchdog' role the *Sapere*
group compiled a list of noxious chemicals in common use, together
with a breakdown into components and brand names – necessary,
since labels often conceal the nature of containers' contents from
shop-floor workers. Such information is essential in opening up
manufacturing processes to democratic scrutiny and is strenuously
resisted by managers where the law permits. In other connections
the radical researchers helped to strengthen bargaining agreements
which dealt with maximum levels of dangerous substances.

After the explosion of the Givaudin-La Roche factory at Seveso,
which contaminated earth, buildings, animals and people with the
toxic compound dioxin, the Montedison research group worked
with factory operatives to reconstruct the events. Early in 1981,
Montedison shut down the whole research group. They are now
being offered golden handshakes to see if they will leave the company quietly. End of limited liberated time. Commencement of
wage-work's alternative – 'free' time, unemployment.

Liberated Time in Dutch Universities
The second example of contemporary 'science for the people' is
much more strongly entrenched in institutional terms. It is the
Dutch phenomenon of 'science shops'. They exist at many universities and technical colleges: Gröningen, Amsterdam, Leiden,
Delft, Nijmegen, Rotterdam, Eindhoven, Maastricht, Triburg,
Wageningen, the Free University of Amsterdam. The science shops
have their origin in the student movement of the early 1970s, when
Dutch students and teachers developed 'project education' as a way
of breaking down the conventional barriers between universities and
the local community, students and curricula, staff and students. For
example, the chemistry department at the State University of
Utrecht had a group which did work for environmental, community
and workers' groups, in answer to questions that they posed out of
struggles in which they were involved. Eventually the rug was pulled
by the university by casting doubt on the scientific quality of this
politically sensitive activity, and the project group set up a chemistry
shop in 1974, outside and with no official support from the university. It still exists, and serves to connect research workers in the

university with groups of people outside who have technical questions that they need to have answered.

Around the same time a number of other department science shops were being set up, mostly in chemistry or physics. But as they became established, and as the questions asked were often far wider than researchers' individual disciplines could cope with, a move began towards shops at university level. Most of those that now exist have the official support of the institution. The role of science shops is not to carry out research but to consult with clients, then advertise the problems around the university and, if no response is forthcoming, to advocate a problem in order to find researchers willing to undertake it. To be acceptable, clients have to satisfy a number of conditions. They must be unable to afford to have the research done elsewhere by a 'straight' agency, they must not have a commercial interest in the problem, and they must be in a position to make practical use of the research results. The questions posed to the science shops vary widely. They may be asked to comment on a questionnaire which an employer is circulating to workers, or to help in analysing a company's accounts. They may be asked to find references to published work on a particular topic and to help the client get access to articles or books through the university or college library. Many questions concern health risks and energy, especially nuclear energy. But they can be more personal. 'Why do people like dogs, and what are the psychological effects on a dog of living in a flat?' They have also taken on and won fights with large companies on behalf of trade unions.

Detached Partisanship
One strong contrast between science shops' work and the consumer research proposed in the 1940s is that because the aim is to assist 'pressure groups' in their struggles many of the enquiries centre on the refusal of 'progress'. Many people using science shops experience themselves as threatened or oppressed and see science, in the normal corporate role, as part of the threat. This means also that science shops tend to be involved more in campaigns than in policy-making. Trade unionists at a small shipyard near Rotterdam had found themselves suffering headaches, sickness, skin rashes and other symptoms after working to repair waste-incinerating ships, and they had failed to get any response from either the company doctor or the factory inspectorate. The chemistry shop at Leiden was able to work with the trade unionists to prepare a questionnaire and interview workers, which led to an inventory of workers' complaints being published as a pamphlet. This helped to give the workers confidence in the issue and to take action to secure safer working conditions. Because of the complexity of the mixtures of wastes there was little chance of a detailed toxicological study; the science shop's help was mostly in organising rather than technical

advice.

Unlike 'consumer research', the science shops are in opposition to institutions of the State apparatus (such as factory inspectors) and manufacturers, and rather than 'discovering' problems which then become defined as needing a technical-bureaucratic solution, they help clients to refine their own definitions of problems and to take action directly to remove them rather than relying simply on administrative action. Science shops involve an alternative (sometimes oppositional) notion of democracy.

The politics of 'discovery' are refused in Britain by, for instance, the Union of Physically Impaired Against Segregation. In their policy statement they write:

> We reject . . . the whole idea of 'experts' and professionals holding forth on how we should accept our disabilities, or giving learned lectures about the 'psychology' of disablement. We already know what it feels like to be poor, isolated, segregated, done good to, stared at, and talked down to – far better than any able-bodied expert. We as a Union are not interested in descriptions of how awful it is to be disabled. What we *are* interested in are ways of *changing our conditions of life*, and thus *overcoming* the disabilities which are imposed on top of our physical impairments by the way society is organised to exclude us. In our view it is only the actual impairment which we must accept; the additional and total unnecessary problems caused by the way we are treated are essentially to be overcome and *not* accepted. We look forward to the day when the army of 'experts' on our social and psychological problems can find more productive work.

While partisan in workers' interests, and while prepared to take oppositional action (in suppressing dangerous goods, for instance), the consumer research institutions proposed by the 1940s' radicals still maintained a clear distance between research and the consumer. A great deal of stress was laid on the conventional guarantee of academic freedom, publication of results. The Consumer Research Council was to publish formal research reports 'which should be in as readable a form as is consistent with scientific integrity', and was also to publicise through press and radio the more important findings about 'widespread habits' such as overcooking of green vegetables which destroys nutritional qualities. It was stressed that good relationships had to be kept with bodies such as Women's Institutes and Cooperative Guilds in order to pass on findings and to glean consumers' problems, and the public had to be educated 'to help them distinguish the false from the true' in goods and advertising.

The whole project allocated a passive role to consumers. Like the consumers of *Which?* magazine, they are not only trapped by the market in what they eat, use and live in but they are also targetted as consumers of researchers' improving knowledge. The proposed

219

Consumer Research Council was to comprise permanent research personnel working on consumer research, supplemented by others experienced in social science research, and with a view to representing the consumer 'further supplemented by a *small* number of lay*men* chosen by the Lord President': Members of Parliament, trade unionists, prominent Cooperators. The people were clearly never seen as real, merely as objects of study and receptacles for knowledge derived elsewhere, by scientific discovery, at telescope distance. The people could only be known statistically, as a distribution of sizes and shapes of feet, to be efficiently fitted with boots and shoes, or patterns of domestic living, to be accommodated in machines for living.

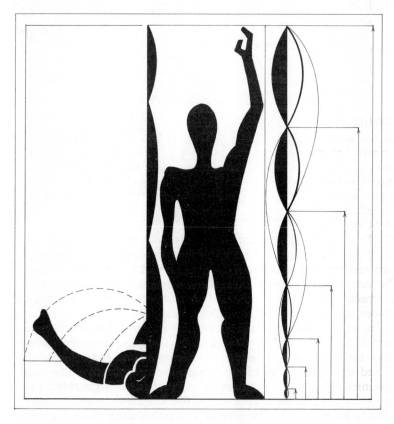

The Modulor 'A harmonic measure to the human scale, universally applicable to architecture and mechanics', invented by the French architect, Le Corbusier, in 1947. The Modulor constitutes a range of standard dimensions to interface people's bodies and the products of mass production. It was also intended to interface designers and their designs: the claim was that 'bad design is made complicated and difficult and good design simple and easy'. Corbusier patented the original Modulor system.

In one pilot study at the Ministry of Works a small number of 'environmental factors' were to be studied in the 'living habits' of a group of about a hundred and fifty families on a number of sites. Each site was to have one observer who would study the remaining tenants – 'artisans and black coated workers' – and also 'together with his wife' acquire first-hand knowledge of the difficulties of finding living space for a family of five in that type of house. One honorary human subject with a personal input to the study, the rest, objects of observation. One day they might read about themselves in the Research Council's papers; or perhaps they would just become happier and happier as the quality of housing improved through the machinations of science. Statistical methods were an intellectual fashion on the 1940s, but it was a 'professional' middle-class stance which determined the distance that was designed into consumer research proposals. Working people were objects of scientific 'popularisation', just as the lower classes had been cast as consumers by middle-class propagators of knowledge in the 1830s.

There is a complicated phenomenon here which needs a name. I call it radical professionalism.

RADICAL PROFESSIONALISM

It has many present day modes. One is 'whistle blowing', which means that when workers in a field see an actual or potential threat they go, not to the self-policing professional community, but to the people who may suffer the damage. This happened, for example, when the City Council in Cambridge, Massachusetts, stopped all genetic engineering research in a university-laden town – Harvard, MIT and many others – until they could be satisfied about its safety. To be successful and to survive in this mode can demand a lot of hard political work from the critical professionals, not to mention the victimisation and anguish they will suffer. Although it goes beyond whistle blowing by being more closely and continuously connected with other workers as a constituency, the Montedison *Sapere* group's fate shows the personal dangers.

In its more continuous and systematic forms 'science for the people' has a very wide political spread. It ranges from the liberal and the pluralist to the committed anti-sexist and anti-racist, Third-worldist and 'workerist'. Liberal radicals engage in advocacy planning, as in American debates over the projected anti-ballistic missile system in the mid-70s. The pluralist idea – that all voices should be heard and that politics is debate – is implicit in many resident's campaigns, as also in campaigns leading to public enquiries over (say) nuclear installations. The now standard procedure of lining up 'our' experts against theirs might be ritualised further, and actually save time without materially affecting the outcome in some cases, by just doing a simple head-count. We win, 11–7, with special weighting for the complex labour power of Nobel Laureates.

221

Feminists and anti-racists contest findings in psychology (IQ) and sociobiology which impugn the rights and status of women and blacks, radical technologists design 'intermediate technology' systems for production in industrialising countries, trade union and community activists engage with managers, administrators and inspectors over industrial and environmental hazards. Activities like these have grown in scale and significance over the past fifteen years. They centre in the radical use of professionals' skills, professionals' know-how and professionals' connections as established members of an intellectual and managerial apparatus.

There are considerable difficulties with this way of doing things. For one thing, most professionals do not really have much power. Even the academic's meal ticket of tenured status is now rapidly being converted into a normal employment contract by monetarist-inspired offensives. When academics drop their conventional 'neutral' status they are likely to be victimised. A liberal Labour science policy minister encouraged the Dutch university science shops, but an education minister has fingered them as a threat to the 'objective and unprejudiced' status of institutions of knowledge. Science shops' low level of overheads (copying, advertising, a storefront, pay for a small staff) gives them room to manoeuvre, so that when a Conservative Minister froze the budget allocation of the Amsterdam University shop the university was able to find an alternative way of feeding it the £17,000 it needs each year. The message for radical professionals, especially in a crisis like the present, and especially in industry, is Don't Stick Your Neck Out, or a cut will land on it. This is the naked political form of the OR Dilemma: to be involved and endangered, or to withdraw from the fray and keep your head down.

Science shops highlight a second contradiction of radical professionalism. The career system of professional work can materially limit the range of work that researchers are willing – not *able* – to do. Clients' projects have to be approved for 'scientific quality' before the university or college can give official support, and bound as they are by the expectations that they will publish results regularly, academics are inhibited from taking on any problem which looks sufficiently real to be technically insoluble or unpublishable. A scientist gets no credit, in the world of 'normal' science, for taking on problems that cannot be solved, however much they matter. Students are caught in the trap too. Unless they can accumulate credits (career capital) for the work they do in 'project education' they resist the idea, and in any case the effect is to narrow the definition of the problem to what suits the student. Precisely the situation of the OR technician, who trims managers' problems to fit the OR tools.

This relates to a third contradiction. Established professionals with career and financial commitments are also committed to specific intellectual approaches and particular ways of working in and out of the lab. Because of these, radical professionals carry at least some of the culture which 'clients' – working people – are up against. Radical professionals do, to at least some degree, want to use what they have been trained in; measurement techniques, abstract forms of reasoning, familiarity with the technical literature, articulate and often patronising styles of debate, formal modes of presenting nice tidy sets of conclusions. In many situations this approach to knowledge can be a liability, not just because it does not work on real, rough, shapeless problems and has to be treated as only one approach among others, but also because in working this way professionals show themselves to be part of the problem and thus raise tensions. Always, the first right of subordinated people is to refuse the culture that carries the dominant groups' hegemony, and it therefore can often be the case that professionals – radical or not – get the cold shoulder from working people in struggle. Just as radical academics in the early Seventies sometimes cast their radicalism in a form of defence of their established interests, and students had first to fight for possession of a territory on which they had always been only guests, so working people may need to challenge the assumptions of radical academics concerning their own roles and the usefulness of what they get paid for knowing. The fact that many of those 1970s students are now the professionals does bring some hope into the situation, but it doesn't ease the institutional contradictions of career and finance.

Experts and Working People
It has been argued in recent years that experts should be seen as a distinct class in industrialised capitalist societies. The term 'Professional–Managerial Class' (PMC) has been suggested, but since 'class' implies a little too much self-awareness and practical coherence, it might be better to carry on using the term 'experts' but with a bit more definition than is usual. 'Experts' might refer to three kinds of workers, all characteristic of 'modern' societies:

1. Those in 'cultural' occupations: teachers, lecturers, social workers and psychiatrists, media professionals and advertising agency staff, designers. Trade union officials too, and MPs. Their work is explicitly concerned with interpreting the world to other people and with developing their general abilities to exploit their social potential.

2. Those in medical occupations: doctors, psychologists, nurses, psychiatrists. Their work is to do with reproducing people in body and mind, in a more mechanical sense.

3. Those in managerial, scientific and technical occupations: engineers, research workers, managers and middle-level administrators. The work they do is directly or indirectly aimed at studying and fixing the work that other workers can do in particular

workplaces.

In contrast with experts, 'working people' are labourers, operatives, craftworkers, clerical workers, service workers and technicians without a college education.

One key characteristic of experts as a class, as a distinct power interest and cultural force, in society is that they are wage workers, they do not own enough private wealth to live off it and therefore they have to go out to work. But the second characteristic is that rather than directly producing things or moving them around – assembling TV sets, erecting buildings, emptying dustbins, serving food – experts fix other people's ability and opportunity to do things. Managers, supervisors and teachers do this by policing work as it happens in factories, offices and schools. Managers, designers, engineers and research workers (and teachers again) do it through fixing the apparatus of production – timing, spacing, machinery, materials, money, processes – and human physiology is maintained as a part of this apparatus by workers in health care. Cultural workers (including, once more, teachers) fix other people's ability to do things by giving shape to their ideas about their place in the world, their role in society, their identity in personal relationships. Experts perpetrate ideological and technical fixes on the autonomy of working people. Their work is hegemony.

Whether 'experts' are also 'working people' is ambiguous, and intentionally so. Some are, some are not. It may vary from hour to hour, year to year, at a personal level, depending on just what kind of occupations we are talking about. Nurses, for instance, are much more 'working people' when with the consultants who lord it over them than they sometimes are with patients. The category of 'experts' is certainly not uniform, but it does help in thinking about cultural and political issues. Experts confront working people on issues around culture, skills and knowledge. Experts suffer, like working people, both as objects of other experts' attentions and as wage workers at the mercy of capital accumulation. The relationship between experts and workers is not one to be resolved intellectually, as a matter of categories, because the roles are firmly entrenched in the structure of modern knowledge and work. The problem will have to be worked out in practice and radical professionalism is one of the arenas of that historic project.

In exploring in this chapter the relationships between consumers and knowledge three key notions have emerged. They form a hierarchy, representing different class-geographical distances between experts and workers. Potentially the closest relationship is in radical professionalism, where partisan experts and working people, as in some of the science shops' work, can move across occupational boundaries to produce knowledge together. It is, however, necessary to be on the alert for the detached partisans who go round 'discovering' other peoples' problems and making 'studies' (that word is a giveaway). The second, more distant relationship is that of watchdog science, which can be exploited by radical professionals as a site for gleaning knowledge but generally serves to take the definition and protection of working people's interests firmly away from the constituency it is supposed to protect. The third relationship – the necessary basic mode – is a negative one; working people's refusal of experts' presumption that they have any right at all to fix the conditions of other people's lives. The challenge of this last situation is to turn refusal into a positive refinement and assertion of working people's own needs and approaches. How?

The Academic Supermarket

THERE ARE ANY NUMBER of academic approaches to science in society. Since 1960 new sub-disciplines have been opened up and new convergences of disciplines have been organised, around problems of science at a number of levels. The management of research and development in industry and the state generated a need which research units for 'science policy' or 'technology policy' have tried to satisfy.

Think-tanks go back further, to explicitly military connections and 'systems' or 'behavioural science' methodologies. A swing from science and counter cultural movements of the late 1960s and early 70s appeared academically in 'science and society' teaching and research groups, which drew on discussions from sociology, history, anthropology, Marxism, cultural criticism and the folklore of natural science.

In institutional terms the courses and teaching units which formed were mostly marginal, appendices of established departments, and in the 1980s many of them face curtailment and closure under budgetary pressures stemming from monetarist State policies. Intellectually, too, many have not really found a home. In negotiating an academic niche, 'science studies' in all its forms has generally argued around standards and traditions stolen or borrowed from mainstream disciplines. The quick-witted were able to play off both ends against the middle for a time, but the question of an authentic intellectual tradition for academic studies of sciences' social relations has generally been fudged. The marginal science studies institutions now face a climate that is financially and ideologically hostile, without either strong ties with conventional academic disciplines or a strong alternative rationale.

If we look to academic work for help in understanding science in practice the most we can do is treat it as supermarket shopping and be fairly eclectic. There are plenty of intellectual traditions to draw on but none of the institutional connections are right from the point of view of non-academics' needs, and this is apparent in the literature. This section lists briefly the main British academic approaches to the kind of questions voiced in this book. Broadly speaking, as the list proceeds the approach gets closer to that of this book, but essentially this book is written from outside of *academic* interests in science. Some of the books listed are awkward to get into. Even so they repay critical reading.

Science and Society

This approach is the easiest to square with actual work in sciences and the self images of scientific workers. When the British Society for Social Responsibility in Science became explicitly radical during the 1970s almost all the members with established careers in natural science resigned and some of them formed a Council for Science and Society. The approach of its members is to refine the common-sense of science and, through exploring why science is so wonderful when left decently alone by outsiders, to make it tick over better: 'Don't worry – I'm a scientist and a responsible one, and I'll tell you if anything's wrong.'

John Ziman, a real scientist and until recently a professor of physics, writes the most popular books out of this stable: *The Force of Knowledge* (CUP, 1976) is one. He conveys a lot about the nuts and bolts of natural science – as a professor sees it. Dr J. F. Ravetz writes at a more abstract philosophical level about why sciences go 'wrong', how craft traditions get 'betrayed', and so on. His *Scientific Knowledge and Its Social Problems* (Penguin, 1973) comes out of such working scientists' concerns but develops them academically rather than politically. Most of the material published for use in schools – for example, material published by the Association for Science Education and by SISCON (Science in Its Social Context) – propagates this kind of refined natural scientists' folklore.

Sociology of Science
Academically the most deeply entrenched, this approach takes the general assumptions of the dominant approach in sociology and applies them in looking at the patterns of organisation and growth in sciences. It has a bedrock of uninspired statistical work, on the numbers of scientific workers and journals, sources and growth of funding, communication networks. Because of sociologists' own hangups about being scientific, the approach looks especially for the Holy Grail of 'objectivity', seeking it in the procedures of control and the norms of professional practice. The problem of whether knowledge is true or not – in general, the problem of *evaluating* science – is left for philosophers to tackle. Sociologists just want the facts: the 'structures' and 'functions' which make up social systems.

Robert K. Merton set this ball rolling in the 1930s and his main papers are collected with an introduction in *The Sociology of Science* (Free Press, 1973). Puritanism and the rise of science is one classic theme on which he worked, rather quantitatively. Arnold Thackray says some interesting things about the social affiliations of early nineteenth century science in 'Natural Knowledge in Cultural Context: The Manchester Model' (*American History Review* 79 (1974), 672–709) but again, the data get in the way of the story.

The Sociology of Knowledge
This approach still steers carefully away from any commitment to evaluating knowledges, which is a little paradoxical since its central concern is to connect the descriptions and evaluations of sciences – their images of the world – with the interests that the knowledges serve. Those who adopt this approach try to reveal where the particular truth of a particular knowledge is rooted, and what it supports, but without venturing to say anything about 'better' or 'worse'. Again, just the facts about the interests being served. But sometimes penetrating studies come out showing, for example, how Mechanics' Institutes propagated a form of scientific knowledge well fitted for particular cultural and political projects of the early nineteenth century, and how physics in Germany after World War I connected with the cultural-political concerns of the Weimar Republic.

These topics are dealt with, respectively, in Steven Shapin and Barry Barnes, 'Science, Nature and Control: Interpreting Mechanics' Institutes' (*Social Studies of Science*, vol. 7 (1977), 31–74), and Paul Forman, 'Weimar Culture, Causality and Quantum Theory, 1918–1927: Adaptation by German Physicists and Mathematicians to a Hostile Intellectual Environment' (*Historical Studies in the Physical Sciences*, vol. III, 1971, 1–115). These titles indicate the extent to which the authors are writing for professional readers who cull their reading from abstracts journals. A book which stimulated much of the 1970s interest was Peter Berger and Thomas Luckman's *Social Construction of Reality* (Penguin, 1971), and a reader of articles from various sociological approaches is Barry Barnes, *Sociology of Science: Selected Readings* (Penguin, 1972). The journal *Social Studies of Science* operates within the general academic area.

Anthropology of Knowledge

Some social anthropologists declined to accept – as sociologists do – that the practices we call scientific are so different from ordinary, non-intellectual practices that they demand a special approach and reverence. The belief systems of tribes can be studied, along with the ways in which their social systems give members a place and a worldview. The same has been done with sciences by authors including Robin Horton, in 'African Traditional Thought and Western Science' (*Africa*, vol. 37, 1967, 50–71 and 155–87), and Mary Douglas, in a number of books including *Rules and Meanings* (Penguin, 1973). David Bloor takes the approach into the forbidden territory of the 'queen of the sciences', pure maths, in 'Polyhedra and the Abominations of Leviticus' (*British Journal of the History of Science*, 1978, 11, 245–72).

These sociological and anthropological approaches address the question: What makes sciences tick? They don't ask who winds them up, and where the keys are. The 'science and society' approach has a simple answer to this: industry and government hold the key, in funding, and these interests point *applied* science in the direction they want it to go; but nobody directs basic research – so hands off! For answers that are more useful to non-professors we need to look elsewhere. Four variants of Marxism have offered models of how to think about power in relation to science and technology, and are linked to various tendencies within radical science.

Base-Superstructure Theories

Beginning with Friedrich Engels and developed by Soviet Marxists, one intellectual approach treats developments in sciences as direct reflections of dominant and emergent economic needs: the 'base' is the economy in a narrow technical sense, the 'superstructure' contains ideas, including scientific ideas. This way of looking at science burst upon British scientists in 1931, when a shining example was set by Boris Hessen in his paper on 'The Social and Economic Roots of Newton's "*Principia*" ' at the International Congress of the History of Science and Technology, published in *Science at the Crossroads* (reprinted by Cass, 1971). British writers who took up the banner in historical studies include J. D. Bernal in *Science in History* (Penguin, 4 vols, illus., 1969), Joseph Needham in *Science and Civilisation in China* (CUP, 7 vols, 1954–) and J. G. Crowther in *The Social Relations of Science* (Macmillan, 1941).

Bernal's writing on science, especially *The Social Function of Science* (Routledge and Kegan Paul, 1939) has had a wide influence outside of Marxism, reaching into 'science policy' studies and merging at some levels with the quantitative approaches of the sociology of science. An illustration of science policy researchers' interests is Christopher Freeman's *Economics of Industrial Innovation* (Penguin, 1974). A left history of science policy is found in Hilary Rose and Steven Rose's *Science and Society* (Penguin, 1969) and they have gathered critiques broadly related to a 'Bernalist' tradition in *The Radicalisation of Science* and *The Political Economy of Science* (Macmillan, 1976).

Critical Theory

Towards the end of his life Engels broadened his understanding of the 'base' to extend to 'the production and reproduction of real life'. Theorists with interests in the politics of culture followed this lead, among them Georg Lukács in *History and Class Consciousness* (Merlin, 1971), Karl Korsch in *Marxism and Philosophy* (New Left Books, 1970) and Antonio Gramsci in his *Prison Notebooks* (Lawrence and Wishart, 1971). Concentrating on what Gramsci called 'hegemony', the Frankfurt school of critical theorists developed a critique of the 'one-dimensional' form of scientific thought. Some examples – essentially pessimistic in their neglect of struggle as a historical process within hegemony – are Herbert Marcuse's *One Dimensional Man* (Sphere, 1968), Jurgen Habermas' 'Technology and Science as "Ideology" ' in *Towards A Rational Society* (Heinemann, 1971; reprinted in Barnes' reader) and Alfred Sohn-Rethel's *Intellectual and Manual Labour* (Macmillan, 1978). This tradition of thought was imported from continental Europe to Britain during the early 1970s but remains more academic than base-superstructure thinking, which links with some self-styled progressive viewpoints within the labour movement.

Labour Process Theory

Re-readings of Marx's *Capital* during the 1970s gave a revived awareness of how capitalist production connects with knowledges in science. Labour process theory analyses science as a process of production, using the same sort of concepts as for other forms of work: raw materials, instruments of production, purposive activity – which are combined to produce a product or use value. Critiques of scientific management came from the USA, notably Harry Braverman's *Labor and Monopoly Capital* (Monthly Review, 1974). These have a pessimistic emphasis on capitalist control which links them with Sohn-Rethel's arguments. Political struggles in Italy and France in the late 1960s produced an emphasis on self-management which appeared in André Gorz's writing, for example, contributions to his edited collection, *The Division of Labour: The Labour*

Process and Class Struggle in Modern Capitalism (Hassocks; Harvester Press, 1976).

Cultural Materialism

British approaches to social and cultural history and – more reluctantly – cultural theory have tantalised members of the radical science movement for a decade. Raymond Williams' writing on literature and society has moved towards Marxism and in the process his ideas have become increasingly effective as ways into thinking about science. His *Marxism and Literature* (OUP, 1977) is dense but repays work; *Culture* (Fontana, 1981) is more approachable. The Radical Science Journal collective seems to have worked its way to here, via critical theory and labour process theory, and argues its approach in contrast to other radical approaches in 'Science, Technology, Medicine and the Socialist Movement' (*Radical Science Journal*, 11, 1981, 3–69).

The British Modern Movement

1930s

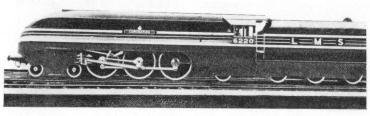

Progress!

WE HAVE TO LET GO our desperate grip on this two-edged commitment to 'science'. As documented in the chapters of this book, the whole project of science as we know it, and especially the definitions of progress which operate through it, is problematic. Somehow we have to learn to face and outface that fact. Sciences, over four centuries, have made a pre-emptive bid for 'rationality' and 'knowledge' as such, so that nowadays we tend to feel that if it's not scientific – quantitative, formalised, distant from perceptible personal concerns, professionalised – then it can't be rational, or real knowledge. A first essential step towards reclaiming this territory as human ground, part of a common culture, is to refuse 'scientific' as an evaluative term. It is only a description, a label applied to certain historically local, interesting, powerful but culturally odd practices. There are more ways of being rigorous and rational than professionalised sciences' ways, and more people who know things than experts. They must be taught firmly that we live here, too. A first step towards a more open culture is to reclaim the air-time from the experts and find voices of our own, vocabularies of our own, with which to speak about knowledges and needs and progress.

There are many possible voices, but this book has five which perhaps speak most strongly. We can characterise them through five terms: alternative products, liberated time, class geography, preconceptualisation and violent science.

'Alternative products' as a focus of discussion leads to an awareness of the differences and conflicts between use value and market value, and then in turn to a growing understanding of how hegemony – the power of financial and managerial elites – is propagated through the fixing of what we make in our work. As a focus of action it is a powerful idea because it centres on what we do, the visible results of our creative activity in the world. It is doubly powerful because it highlights 'the facts' of material existence – the things we live and work with – and questions just how necessary they really are in the form that they presently take. Perhaps the one idea from progressive images of science which we might build on is the notion of 'the freedom of necessity', the power which comes from knowing what has to be the way that it is, physically, culturally and historically. Probing reality through the investigation of alternative products in all areas of work – factories, offices, hospitals, schools and colleges, homes – will refine our sense of how much of what we live with really is necessary.

'Liberated time' has a similar power, focussing not on the products of work but on the working itself. If our needs are to be refined and asserted then this must be not only in debates and negotiations but also in the day to day ways in which we connect up with the forces of production in society. Our values must be made visible in the time we spend doing the work that we do or could do, using the tools and skills that we use. We might reclaim some of our own work, not for privatised use but collectively, to define and create something which we know directly satisfies an expressed need. Not something sneaked for ourselves, or for the firm to sell; although they might eventually sell the product, liberated time is essentially a challenge to both individualistic and market values. Not something that the conventions say we should deliver. But something that expresses our support for other people in their struggles to live and make progress.

'Class geography' is an idea which tries to grasp how some people's actions are central in directing progress and others' are peripheral. We experience this in our own thoughts, when we find it hard to know what we think. We see it at work, when it's made obvious that we're not paid to think. We see it in local communities and national politics when the money that is spent and the commitments of resources that are made cut right across any scheme for which we can see the sense. We see it at an international level, when Third World economies and Third World peoples are held in thrall to First World financial interests and political elites. The idea of class geography tries to grasp the relations of economic and political power between central and marginal groups, metropolitan and peripheral countries. But it also embraces the senses of identity – potency and powerlessness, superiority and inferiority – that we all feel in some social locations, and feel differently in other locations. It tries to focus consciousness on the 'uneven development' of both power and feelings of power, by probing how the general power relations of production are materialised in localised practices within the forces of production, and how they shift over time taking with them power, prestige and personal identity.

'Preconceptualisation' identifies a 'fact of life' of so-called scientific ways of organising society, including sciences themselves. Some people design, others make. Some people think, others do. Some people have access to the means of thinking; libraries, bodies of knowledge, information technology, design tools, time to think, places in which to meet for discussing and planning, communications networks. Other people are materially excluded from use by physical barriers, elitist discrimination, legal injunctions, cultural taboos, task boundaries, numeracy, literacy. Some people determine what goods and services and knowledges are to be produced, and others if they get access at all, merely consume. Preconceptualisation is a powerful concept for generating critiques and alternatives, because the relationships that it names are so deep and fundamental in the organisation of industrialised life, and so profoundly anti-democratic. Its sweep includes not just 'the separation of conception and execution' in industry and office, but also the whole systematic abstraction of knowledge which is science.

'Violent science', finally, points to the fundamentally problematic nature of the emotional project implicit in and reproduced through hard, reductionist science. It highlights the systematic discrimination against and suppression of personal insights and responses, the repression of emotionality and needs other than the privileged emotionality and needs of science: cool, distanced, assertive, manipulative and control-seeking. Preconceptualisation stresses the political importance of production, and class geography shifts attention to the ways in which production is also reproduction of power. Liberated time and alternative products point up specific ways in which politics must be cast as the literal re-production of a social order. But then what violent science points to is something beyond all of these, taken narrowly as they might be. The masculinist ethos of science and industrial society tends to make even reproduction – the definitive female function – into a sub-category, a subordinate form of production, a function which male culture has appropriated as definitively masculine. While these definitions cannot be accepted as necessary, and the cultural ground must be reclaimed, they are difficult to oppose. What 'violent science' tries to do, as an emergent critical concept, is to direct attention to the potential power of nurture, rather than reproduction, rather than personal development, rather than simply 'needs', as an idea and a form of human action through which to counter the violence of science and its social order.

When worked together all of these voices, and the forms of

236

practice that they grow from and inform, potentially define an alternative mode of knowledging reality. By linking fact, fiction – that is, the imaginative projection of alternatives and the sensitive capturing of class-geographical 'facts' – and action, they tend towards a practical definition of something we might call *faction*. As a way of knowledging the world and thus a way of reading and writing reality, faction *is* factional; it is partisan, intimately connected with specific people and their needs and places in society. But clearly, from this book's criticisms of the elitism and partiality of sciences, this need not necessarily mean that faction yields knowledge any the less rigorous or true than the sciences do. What must be sought through faction is the freedom of knowing what is and is not necessary, and a form of social organisation among working people which can produce knowledges with clear class relations in practice and rigour sufficient to meet the most exacting practical demands. This means the established empirical grasp of natural science; but it also means much more. Especially, it demands that producers of knowledge, working people and experts, learn to study not only their objects of research but also where they stand. Fact, fiction and action, transactional, poetic and expressive ways of languaging, all must be worked together to achieve this new standard of knowledge. Faction is the mode of people's research.

New knowledges only come to exist as new practices are created. Finding voices to speak with is therefore the same as making places in class geography to speak and be heard and act in. Schools, colleges, welfare services, political parties or trade unions are not terribly good at doing this. And it is something that corporations, police and armed forces, and governments oppose. It threatens 'order'. But new knowledges, sciences' successors tomorrow, require not just new definitions of progress and new definitions of knowledge but new definitions of order too.

12 Living Well: Democracy and Science

THINGS HAVE COME ON A LOT since I was a kid. I grew up in a privately rented house on a cobbled street, one room up, one room down. Gas stove in the cellar-head, stone slab sink in the corner cupboard with one cold tap, bathing in a tin tub of water heated in the cellar wash-boiler. Each room had a window looking straight across to the identical houses opposite, blackened Pennine sandstone; and they backed on to an identical row, which faced an identical row, and so on. Washing strung across the streets was hand mangled in the cellar. The WCs were outside, three shared between six houses, next to the midden where kids played bandits and climbed on the roof. I was living here, aged five, when the Festival of Britain blared forth its message of modernism.

All those streets have gone now, bulldozed during the sixties. A system-built all-window flat-roofed school stands there surrounded by grass – grass! – in place of the massive church-like Board School,

238

its squeaking chalk and echoing high beams, head-high windowsills and twelve-foot perimeter wall.

The two-bedroom Council house that we moved into opened up all sorts of possibilities in the 'You've never had it so good' post-war boom. When you have a bathroom you can buy . . . soap racks, wall cabinets, wallpaper, lampshades, toilet-seat covers, toilet-seat cover cosies, coloured toilet roll, soap to match, carpet; endless possibilities. The kitchen was simple but it was a kitchen. Double drainers, fitted cupboards, electric sockets and enough room, eventually, for a washing machine and a fridge to stand on the lino-tiled floor, and stainless steel pans to hang by the food blender. The windows – front and back – looked out over gardens, to be stocked over time with flowers, shrubs, trees; fertilised, insecticided. Gradually the Festival of Britain came to live in our house, and the houses of everybody I knew. Flush doors and ball catches, black and red 'molecule' magazine racks and Jackson Pollock stick-on vinyl all came and went. But the apparatus and the technique came to stay. The need to fill out the empty ideal home-shell worked on nights and Sundays, it worked each holiday, it settled down among us and it never went away.

The tale still goes on. I left home, on a State grant to university, and bought an electric guitar. I left university, as a graduate engineering trainee, and bought a hi-fi, a record collection and a car – two cars. I became an owner-occupier and the house got liberally rewired with double sockets and fitted with central heating and now, trying to earn a living as a freelance writer and researcher, I itch for the use of a word processor. Phone, car, television, deep-freeze? Of course.

I don't mean to imply that everyone nowadays has all these things. A capitalist society is a heap, and there are always people pressed under by the weight of the heap-dwellers higher up. What I mean to stress is the way that, for a single working-class family over a single generation, the technologically expressed dreams of the 1930s and 40s have become part of the structure of everyday life and expectation and spending. The average domestic kitchen is visibly the offspring of ideal kitchens planned by experts like Lillian Gilbreth, who married and collaborated with Frank Gilbreth, a leader in time-and-motion study. Lillian Gilbreth advised gas companies of the 1930s on how to redesign appliances so that they could be mass produced, and so that kitchens could be presented to women as rationally organised workplaces needing functionally designed, modular, specialised apparatus.

The point of this emphasis is to lead us to wonder what is brewing now. The gas appliance companies and the Lillian Gilbreths have had their day, not to mention organising ours. So whose day is tomorrow?

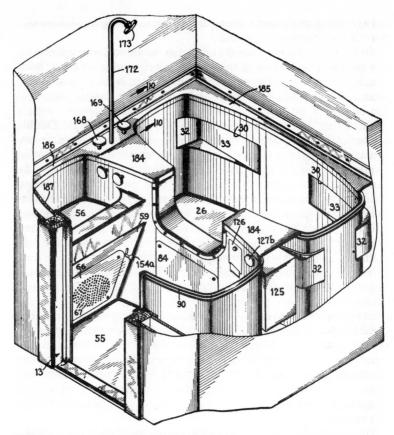

Prefabricated bathroom sectioned in two parts, R. Buckminster Fuller, 1938.
All components are pressed simultaneously with the metal skin, every
square inch carefully calculated. (US Patent 2,220,482) Fuller later designed
an integrated bathroom-kitchen-heat-light unit, transportable in one module
for installation on site as a 'mechanical core'.

Tomorrow's World Today

In London three-quarters of a century ago, in 1908, the press baron
Lord Northcliffe founded the Ideal Home Exhibition, 'a demonstration of the best and latest products for the home-maker'. Something like 900,000 people visit it in a month nowadays. What more
convincing evidence of the pull of modernism – the expression of
science in everyday consumption – and the grip that mass production and mass-produced technology now holds over styles of living?
In the present recession promoters still expect to sell. Even with
three million unemployed in Britain those in work have money to
spend and dreams to build. It was so in the Depression years too.
In 1929 an 'electrical fountain' threw water sixty feet in the air, and

240

in 1932 the biggest, most spectacular exhibition ever drew 700,000 to a 'City of Light'.

No longer bread and circuses but gadgets and circuses, commodities and circuses; technology as a circus. The technological spectacle promises tomorrow today and the wildest fantasies fulfilled through pressing buttons and spending wages. What more striking illustration of the pivotal role of science and technology in modern capitalism? Electricity in the home was a fantasy this side of the rainbow in the 1930s. Electrical technology was the light at the end of capital's tunnel, which together with the automobile was to pull corporations and us through into the world of General Motors and General Electric, leaving behind the world of the General Strike.

Living well is a cultural norm now in 'developed' (capitalist) societies. But who decides if you're living better? You may choose which commodities to buy, to the extent that your income allows you any choice. But who chooses which use values to offer, as commodities? They are the ones who decide whether a possible development would be 'progress' or not. Who chooses which knowledges to produce, to become part of the cultural stock? Profit-interested boards and top managers of corporations, control-interested managers and politicians, career and status-interested directors of R&D, technically interested researchers and engineers: they make the decisions. What kind of cultural apparatus is that? Could its membership be radically expanded to include all of us, at the stages before commitments of money and work get too weighty and the stakes too high? Why not? Could the criteria for deciding if it's progress be made different, open, and open to criticism? Why not? Can we move on these fronts?

The reason for making this book was to argue that moves like this are necessary, and that we can and must begin to make them now. Experts and governments and corporations impose definitions of progress when they call a technical fix in high yield seeds a 'revolution'. They call for 'progress' in installing cable TV links so that new markets can be fed into our homes. They demand funding so that progress in basic research can continue under the supervision of elites, even though it is basic to all our futures. They attack 'progressive' methods in education in the name of efficiency and the scientific establishment. They design jobs, machines, medicines and techniques which limit – and are designed to limit – freedom of choice in the way that they are incorporated into our lives and work. They ask us to trust them as watchdogs and professionals even when outcomes of their work are a manifest violation of bodies, identities and freedoms.

This book has tried to show that there are *no* established bases for trust, and no universally acceptable definitions of progress through or in sciences. Our only guarantees must lie in our own participation in the setting of sciences' goals and the evaluation of their outcomes. 'Truth' is no guarantee, because there is no guarantee of truth except constant vigilance, and those who run science clearly have tunnel vision in what they watch out for. We need guarantees *now*. Sciences, technologies and medical practices are complex and often difficult to see, which is why much of this book has been concerned with understanding histories which confront us today. But the point is to change them, today and tomorrow, and the understandings have been constructed with changes in mind.

As business leaders sit wating for Godot – 'the upturn' – these are the areas in the technological economy that they watch, looking for chances: robots and satellites, bugs and spares, windmills and nodules. In the Eighties these will provide whatever bases there are for accelerated capital accumulation, and there is therefore bound to be 'progress' in these sectors. What we must do, as working people, is to learn to review what has been learned in past struggles around science, technology and medicine, in order to be clearer about the *places* we might move out from and the *voices* we might raise in exploring and asserting our needs against capital's. In earlier chapters we have developed some intellectual aids for this purpose, some means through which we might get to know knowledge better. *Places* can be assessed in terms of the distances they place between experts and working people: refusal, watchdog science, radical professionalism, people's research. *Voices* are many, but this book has offered five main critical concepts to open up debates: violent science, preconceptualisation, class geography, liberated time, alternative products. As for defining our needs that is, of course, the hub of the problem. We can manage, to begin, with a very general kind of definition in opposition to capital's: as aspects of living well we can talk about feeling well, eating well, feeling free, knowing what to do, and being free to do it. These terms of everyday life and work can take us quite a long way.

Three particular areas of life will be looked at in this chapter – jobs, doctors and eating – and also the more general organisatonal issue of campaigns and watchdogs.

JOBS REVISITED

In the last ten years some important new directions have emerged in workers' struggles to protect their jobs, the importance lying in the way they have addressed three basic human needs; feeling free, knowing what to do, and being free to do it. When workers are continually threatened by redundancy or by deskilling through new technology and new work methods they cannot feel free. There is constant worry, mistrust, a legitimate sense of oppression and an understandable resistance to change. When white-collar workers are separated from shop-floor workers and each other, and control the processing of knowledge within the firm, it is hard for any section of the workforce to know enough to make proper plans to change the situation. And when investment and information is controlled by company boards with the close support of the state it is hard for workers to do much anyway. Some new attacks on these 'facts of life' have been developed in the last ten years and many of them can be illustrated in the campaigns of the Lucas Aerospace shop stewards' Combine Committee.

In their struggles over jobs the stewards quickly realised that they had to learn more about the company. One thing which they learnt was to ask questions about new technology and what it would do to the craft skill content of jobs and the numbers of skilled, unskilled and semi-skilled jobs. Would the new machine or process:

□measure or time workers' work or log their performance?

□log how workers managed their work, so that managers might automate it at a later stage?

□pace the worker and fix the worker's way of working, rather than respond to an individual's style and particular know-how?

□exclude an old skill without creating an equivalent new one? There is no technical reason why any of these need happen, and every reason why workers should refuse to let them pass. In their campaigns to prevent redundancy, workers in Lucas Aerospace developed ways of negotiating round these questions, asserting their values in opposition to conventional management values. Preconceptualisation and class geography are the fundamental issues here, two aspects of the socialisation of work. The former involves taking thinking out of the job. The latter involves destroying the established skills which are a central part of workers' culture, and with them probably their pride in their ability to work and certainly some of their uniqueness in a competitive labour market. It makes them more marginal, both in identity and in the market.

Alternative Products

The campaign's focus on preconceptualisation and class geography was not only defensive. The Combine produced a large six volume plan for alternative production. They stress that it was not an alternative plan because there was no publicly available plan developed by the company's managers from which workers could get any idea at all of what the future might hold for them. They set out to design a future in terms of jobs, on the basis of what they knew, collectively, about the company's plants, processes and markets and the workers' interests and skills. At one level they were concerned to protect and enhance their skills while creating new jobs to replace the old ones made redundant by changes in markets, products and techniques. Rather than simply resisting change they figured out changes which were in their interests as workers. At another level, though, they went beyond their narrow interests as producers. They considered the relationship between Lucas Aerospace products and society, and the uses of high-tech processes in the face of the many

243

un-met needs that might be addressed: housing, medical equipment, transport. This led them to widen their outlook and campaign for 'socially useful production'.

The gambit is risky, ideologically, because it is possible to counter it – as Lucas managers did – with arguments that the warplanes which Lucas equipped also satisfied a 'social need'. But the Combine stewards were not talking about a 'need' of an abstract thing called a nation, but a need refined and identified by a specific, visible community of working people as *their* need.

Alternative products mean new standards, being selective, weighing up gains and losses. Not deferring to the present, but confronting the past with a view to the future. The demand to work on 'socially useful' products is a necessary and productive one because it focuses attention on use values, on the actual relationships between working people at both 'ends' of the production process which shapes society. Some of the alternative values can be listed. Products should augment skills rather than destroy them for other workers; sophisticated robot technology could be used to replace humans' presence in arduous conditions, yet still retain humans' skills by having the robots remotely controlled. Products could ease the condition of ill or disabled people, so Aerospace workers designed a 'hob cart' which would help disabled children to feel free of some of the constraints on their mobility. Products could help people in rich and poor countries to feel more free; remote districts would be less cut off if the Lucas workers' designs for a road-rail vehicle were put into production. Products should be energy-conserving and minimally polluting, which is why a hybrid petrol-electrical power pack was designed. The concept of violent science is important here, for implicit in these alternative products is the assumption that workers do not have the right even in protection of their jobs, to assist in the coercion and violation of other workers' lives.

Liberated Time

One of the most radical aspects of the workers' plan was the way in which it was developed. It involved participation across types and classes of jobs, unions and sites between shop-floor and design-office workers. It involved participation across the boundaries of the company; with Health Service shop stewards over the design and procurement of kidney machines, and with academic research workers in universities and polytechnics. Clearly, class geography is again an issue here, especially in terms of the politics of radical professionalism, both within the Lucas unions in terms of relationships of trust, access and specialist language between office and factory workers, and with outsiders such as academics. The 'naturalness' of the expert/worker relationship was – not surprisingly – never fully demystified and overcome. The radical core of the plan, as a form of research, lay in liberated time which is the appropriation of paid wage-work hours for workers' collective purposes, in company time, using company facilities; not just meeting rooms, but machines and information systems. Unofficially there must have been many hours of Lucas work time taken up in developing the plan. But there are clear limits on this since shop-floor workers would generally find it difficult to leave their work or their workplace for discussions with other workers. This can only be a guess but probably the most significant component of liberated time, as opposed to 'spare time' which trade-union activists habitually give up, was that of academics acting either as coordinators of the planning process or research and development workers in Combine-instigated projects. All the contradictions of radical professionalism which we noted in discussing the Dutch science shops must have been active here too, but nevertheless the Lucas Aerospace stewards' Corporate Plan is an ambitious and impressive example of how liberated (academics') time can be incorporated into people's research.

Strategies in Knowledge

Collaboration between industrial workers and non-industrial researchers has been a notable feature of some struggles over jobs in Britain: GEC, Chrysler, Vickers and others, who also developed workers' plans. In many industrial towns and cities action groups have developed into trade union and community research centres funded partly by state money. Researchers based both in these groups and in university or college departments, have built up good contacts and new practical skills in connecting research and action. Through connections like these, trade-union activists are finding it possible to articulate and publish their views on industrial and social policy much more widely. Stewards in the motor industry, for example, published *A Workers' Inquiry into the Motor Industry*, and four Trades Councils published *State Intervention in Industry*. There are two things worth noting about innovations such as these.

The first is the way that the State is critically implicated in workers' struggles. For Lucas Aerospace the state, as procurer of military equipment, was a direct supplier or destroyer of jobs, and Ministers were unreliable allies in the campaigns around the Plan. The workers' plan was not rejected because the proposed alternative products were unmarketable or even unprofitable. It was rejected, first, because managers clung to their sole 'right' to manage, and second, because a shift from war products would have lost the company preferential access to state concessions and capital. For motors workers, the state was heavily committed to plans for 'rationalising' the industry according to manager's priorities rather than workers', market values rather than use values. The Trades Councils' book began as an inquiry into the state rationalisation agency, the National Enterprise Board, and into why things went 'wrong' under Labour as well as Conservative governments. This problematic role of the state has arisen before in discussing consumer research and education in the previous two chapters, and it appears again later in this one. However, it should also be noted that the state at local level has to be regarded in a different light to the national state. In London the Greater London Council has an economic planning department – the London Enterprise Board – within which policies inspired by Lucas Aerospace workers' and similar campaigns are being pursued. The West Midlands council has sponsored a co-ops research unit, which has alternative ways of organising work as part of its remit. Different spaces within the metropolitan-provincial relationships of the state have different values for alternative and opposition practice.

The second thing to note is the emergence of some distinctive forms of organising in people's research. One is what might be called the *learning exchange*. Trade-union activists call meetings, not only to process resolutions and swop facts and horror stories, but also to plan, criticise, and listen to the views and analyses, both of other workers and of 'outside' researchers who are working with them as partisans. They accumulate an informal apparatus of information gathering and distribution, and can use it to find people who may be able to help them do the things which they cannot see how to manage themselves. In its conscious focus on research this is an extension beyond traditional trade-union practice, including the conventional assumption that the leaders and officials of unions are the only ones who have any need of research support. A second approach used is *the audit*. Combine stewards audited the production facilities and skills of Lucas Aerospace workplaces as a basis first for self-education and then for planning. This is an approach that can be widely used, and not only in industry. Adult students in an evening class audited the industrial and social structure of Leeds to produce an alternative city plan. A third approach is *the inquiry*. Its exact form depends on how much money or free time is available. Using the central device of a questionnaire an inquiry can move into interviews loosely built round the answers and into 'hearings' where groups can discuss differences and ask for clarifications. The questionnaire can be much more than a bare form, becoming more like a pamphlet as it provides points of view for readers to bounce off when responding. And of course the findings are reported back to those who contributed, before wider publication.

It has to be stressed that the point of these information gathering activities is not information. It is to build up an apparatus, an organisation, through which workers' thoughts and workers' actions can be made more effective and coherent.

DOCTORS REVISITED

In some ways the British feminist magazine *Spare Rib* functions like the *Which?* magazine of the Consumers' Association. *Which?* offers advice and information on the technical qualities and value for money of commodities. *Spare Rib* – among other functions – offers a forum for information on medicine and medical techniques as they most immediately affect women, especially in midwifery, gynaecology and birth control. The differences, however, are significant. *Which?* stays within the market and the social relations which give it shape, offers 'best buys' rather than self-help, and challenges technical specifications or manufacturing quality control rather than the needs which they are matched against in market terms. It is watchdog science in classic form, and radicals of the 1930s would probably be proud of it.

Although *Spare Rib* itself is a commodity, available on the open market rather than to subscribers only (like *Which?*), this is probably the limit of its essential connection with the values of consumerism. The magazine gives voice, as do other feminist journals, to the 'consumers' of medical specialists' services and drug companies' products. This is central in that it is not only experts' reports which appear in *Spare Rib*, but the experiences and accumulated knowledge of women speaking as the people whose need to feel well should be satisfied. Contributors give guidance on choice within the range: injectable contraceptives versus oral versus intra-uterine, home childbirth and varied delivery positions rather than hospitalisation and standardisation. But they also challenge the limits of that range, and the social relations – most often patriarchy, but also commodity production – which are seen to set those limits.

Alternative products are thus only one focus in arguments about women's needs for medicines and therapies which lie outside the capitalist market and its sophisticated products. Violent science is also very much a woman's issue, since the mechanistic, dominating, hard and distanced – phallocratic – approach of science manifests itself powerfully in male doctors' treatment of women patients. Women have taken up the struggle (feeling free) against the patriarchal relations around which modern medicine is structured. And so class geography is an issue too, since many of the conditions of women's treatment are also those of working people and people of other nationalities, in opposition to experts.

Women's movement traditions are central in this last issue because although there is deep opposition to experts this does not generally slide into a refusal of knowledge, maybe because many feminists are themselves trained for professions. A major characteristic of the movement has been a search for alternative forms of knowledge based in personal experience and the sharing of experiences. On one hand this underlies a whole system of women's cultural institutions, from consciousness-raising and therapy groups, through well-women clinics and self-help medical groups (such as the Boston Women's Health Book Collective, who have published

247

two best-selling people's readers, *Our Bodies, Ourselves* and *Ourselves and Our Children*), to local and national information centres and networks and conferences. *Spare Rib* is thus a popular phenomenon, not a market phenomenon, an expression in the market of a real and active cultural movement. In addition to the variety of groups, however, there is the definitive style of feminist practice – non-hierarchical collective work, giving place to personal knowledge, sharing it and respecting an exploratory tentative approach to expressing needs. The style is tenuous because it is the style of an emergent, marginal culture, and it is contradictory within the movement itself. But together with the actual organisations of the women's movement it provides a model of people's research and of faction as a means of exploring alternative ways of living.

Local Research in Health

The forms of organising which we noted in workers' jobs campaigns are effective in local community-based campaigns too, including campaigns on health-related issues. The basic form is the learning exchange, that of simply getting people together or giving them the means to contact one another in order to discuss specific issues and pool or refine their understandings. Once an organisation of that kind begins to get off the ground more ambitious forms of researching can follow: audits of the uses of land on an estate (play space, shops, health services, houses, factories) or the environmental conditions across a part of a town, perhaps an inquiry into the forms of ill health in a community and their causes.

In any of these activities it could be useful or essential to have the help of health service or social service officials like community physicians, GPs, health education officers, health visitors, environmental health officers, workers on the Community Health Council, who could all be valuable members of a neighbourhood team. Such radical professionals could help with data alternative to that which is normally available (advice from a social statistician is always a good idea when designing surveys or questionnaires). They could ensure the technical correctness of alternative arguments and might, because of their work connections, be able actually to put into practice some changes in organisation and provision. Because of the glass-geography of the health service, however, there are unlikely to be many activities of this kind which do not get hung up at some stage or other on the contradictions of radical professionalism, either by it sliding back into detached partisanship (where the experts forget whose life it is they are 'discovering') or by coming up against the professions' internal police systems.

For instance you may never have heard of an Ectron Duopulse Mark Two, a Transpsycon ZU/ZUSS or a Phasotron. They are not something out of Buck Rogers or Flash Gordon; they are electric shock machines used in mental hospitals for ECT (Electro-Convul-

248

sive Therapy) treatment. Many people find this aspect of violent science repulsive, as did one nursing student during the course of his training. Because he expressed his moral opposition to this kind of treatment he was barred from sitting for his exams as a Registered Mental Nurse and given a month to leave the college. 'I have received support from nurses in other parts of the country,' he said, 'but many are understandably too worried about their careers to take a similar stand.'

Although many of the issues of technological investment are pitched at the state level in the design of hospital accommodation, selection of treatments, funding of research programmes and purchasing of equipment, some things can be done at a local level. For instance, microcomputers are becoming widely used by doctors and they could if they were asked make a useful community resource out of doctors' medical records, opening them up – with proper safeguards for confidentiality – as statistical sources for people's research. Why not?

EATING REVISITED

Bugs and robots are the key sectors here. Bug technology is providing the technical and economic basis for multinational firms' move into the ownership, centralisation and genetic manipulation of seeds, so that the prospects of what may be grown and eaten, where and under what conditions, are increasingly threatened by monopolisation. Who decides if it's progress in agriculture? Initially, the corporate elites and technician elites. By the time that farmers discover that the old varieties of grain were more robust, cheaper to grow and had more uses, and by the time that eaters discover that the old plant products were tastier and had more nutritional value despite the 'wonder' tag on the seed or the supermarket packs – by that time the old seeds may have gone out of stock for good, the old markets may have been absorbed totally by the new, and the processing facilities may have been capitalised to such an extent that nothing short of a revolution would get manufacturers to produce anything else. Protests then are likely to result only in further sophistication through synthetic taste agents and additional nutrients.

Definitions of eating well also come under pressure from robots. It was premature to conclude, at the start of this chapter, that the gas appliance companies and Lillian Gilbreth had had their day. The cult of efficiency and consumer styles continue to smooth the way of consumer goods such as slow cookers and microwave ovens into the kitchen. The pressures of time economy in domestic labour provide very direct reasons for using such gadgets – though they do not save anyone's labour, except by killing off one market or firm to provide a market for another. The main impact of innovation in 'robots' is in the factory. Agricapitalists can use advanced production technology to carry out hyper-processing on cheap food stocks, cheap because research has made yields 'better' and 'better' until too much is being produced in some markets. The result is 'convenience' foods which require a minimal amount of finishing in the home kitchen. Processing in vast quantities, purchasing at low prices, and continually driving labour out of factory processes through automation (bagging crisps, checking the colour of peas) means that manufacturers supply 'convenience' to us at the expense of jobs, nutritive qualities, commitment of families' incomes to manufactured commodities, and further entrenchment of corporate domination over farming practices and eating habits throughout the world.

This may all sound at a very high level, and it is. But the social processes are not out of our reach even in everyday terms. One good case is the investigation carried out by the Lancashire School Meals Campaign into the economics and politics of shifts that are taking place under a monetarist government in providing food for children at school. They found that school meals staff are losing jobs because of rising prices for meals, which cause families to send children to school with packed lunches instead. School-prepared food is being driven out by factory-prepared food in children's lunch packs, in shops near schools, in fast food sold in vending machines in schools or finished in school kitchens. Nutritional values are being destroyed by factory processing. On the other hand waste is increasing ('We used to have fifteen pounds. But forty children come with sandwiches from home now, and we have a bucket of waste left over each day just from them.' 'It's all wrappers, and tin cans and polythene . . . It actually takes us longer to clear up now.'). The burden of cost is being shifted to the private household from the state where great economies could be made in bulk purchasing as in the army; the accounting system there makes the meals 'cheaper' too. Hazards in kitchen work are increasing too, as microwave ovens enter the fray and British safety standards on these are the weakest when compared with other countries'. The Lancashire campaign proposed actions to counter these systematic changes.

Though the food multinationals are an international system and must be tackled as such, the case of school meals shows how much can be seen and acted on where we live and work, in our local communities and local workplaces. The way to tackle food at this level, and to begin to pursue it at higher levels, is to treat it as work. Food is produced by a whole chain of people, some on farms, some in factories and public kitchens, some in homes. What is involved in today's 'progress' is a recomposition of this work as a whole, shifting the balance from one country to another as research input increases, for example, and from one location to another, within a country or a single community. Just as in approaching conventional industrial work, the critical concepts can be used to give voice:

250

preconceptualisation, class geography, liberated time, alternative products – and some pretty violent science too in the breaking down and reformulating of nutritional values.

Although they were not initially conscious of it, the Lucas Aerospace workers' struggle to organise across sites in the UK was a response to the strategies of a multinational company at a time when multinationals were growing rapidly in numbers and power. In its turn, the national campaign has captured the international imagination of people looking for ways of countering the degrading forces of multinational production; more concrete developments have followed from the workers' plan in other countries than they have in Britain.

In relation to multinational developments in technologies and the division of labour we can place ourselves in the same range of places in the sphere of innovation: refusal, watchdog science, radical professionalism, people's research. We can use the same range of tactics: information exchanges, audits, people's inquiries. Over the coming decade we can be sure of further volcanic rumbles from the hot zones of innovation, especially robots and bugs. But we can also – if we choose and if we organise – figure out what is happening and move it in other directions.

WATCHDOGS AND CAMPAIGNS

There are at least two things about campaigns to be wary of. The first is that campaigns around 'technical' issues get bogged down in technical questions, and the second is that campaigns about 'science policy' (industrial policy, economic strategy, etc.) tend to be essentially quantitative rather than qualitative – more, or less, but not different.

On the first point, campaigns against nukes as a basis for the national power-generating system, are a useful example. The crucial point to grasp is that nukes are not about energy, or technology. The argument is not basically about cheapness, cleanness or safety, though these are sideshows set up by groups who want to stay ahead of the game and thus entrench their interests in the growth of the nuclear industry and its political-economic power. Politicians, civil service administrators, military heads, corporations (especially in construction industries) and technicians all can have distinct but converging interests in the power of nukes. That power has very little to do with thermodynamics and nuclear physics. It is political power.

Over recent decades opponents of nuclear power's power have been forced to counter arguments at the technical level since that was the game being set up by the industry's experts. In consequence the 'pro' arguments have all been shown to be technically questionable and the blithe assurances of thirty years ago so much PR hogwash. The experts' claims have got weaker and weaker as real

opposition has eroded the ring of righteous confidence. But corporations want to sell reactors in world markets even if it is obvious that their use in 'friendly' or unstable, fascist or democratic countries will be for producing the fissionable material for nuclear weapons. National leaders want to undercut the traditional political power of sectors of the workforce that are strongly organised or have strong material positions for bargaining – workers in coal mining and the electricity distribution system. Workers in the nuclear industry can be fenced in with an apparatus of 'security' that denies them rights such as sanctions against the employer in pursuit of an industrial dispute, which other workers are able to retain in the face of state offensives. For the state then, as smoother of the way for capital accumulation, nuclear power is an attractive avenue of development in labour control – not energy technology. Nuclear technology is in fact the natural energy choice of the authoritarian 'strong state'. Controls, prohibitions and surveillance extend out from the workplace to the local community, homes and associated industries like transport, for instance, so that whether they 'work' or not in thermodynamic and money terms, nukes are progress for the interests which sponsor research and development in other areas of manifestly violent science; warfighting and civil war technology, for the armed forces, the regular police and the 'third force' of crowd-control police.

A second point about campaigns relates to issues such as proposed alternative economic strategies or alternative science and technology policies. The dominant approach to such issues is the 'progressive' one, which sees progress as given in quality and having its own internal logic, so that political issues have to be framed in terms of uses versus abuses of a 'thing': science. This is fundamentally a consumerist approach and as such suffers from all the weaknesses outlined in discussing consumer research in an earlier chapter. The progressive view is in fact a historical residue of the era in which consumerism became both a successful capitalist strategy and a campaigning platform for radicals. Arguments within this kind of framework are about slices of a cake. They should be about ingredients and ways of working them, alternatives in products and the effects of eating them. 'Planning', the panacea for all ills diagnosed by progressive, expert minded people, is just a form of watchdogism. How many rest breaks, how many cars? How many trained workers, how many loyal party members in key managerial positions? Important tactical questions; but without questioning pre-conceptualisation and the role of experts, policies based on these approaches can only produce change without change, more but not better. How could they lead to better living if they accept that progress is defined outside of culture, technically?

Watchdogs Revisited

Having argued that watchdog institutions are very limited in their usefulness to us, this point needs some further consideration. 'Watchdogs' refers to bodies with official status, resulting either from public campaigning and legislation (like the Factories Inspectorate), government appointment (like GMAG, the British genetic manipulation advisory group) or formal negotiation (like a management/union safety or planning committee, set up to police a negotiated agreement). We may set up institutions to survey and police developments in a particular field. (The organisation *State Research* is an example mentioned in the addresses list at the back of this book; shop stewards' safety committees are another.) But these are not watchdog institutions in the present sense, because they are on our ground and in principle can easily be arms of people's research under our control. Watchdog bodies are 'their' ground. The Factories Inspectorate is experts' ground, and its practice shows that much of the time it is employers' ground too. Bits of ground within are ours, but the institution as a whole is essentially a site of struggle. It may be the location of some radical professionalism but is subject to all the contradictions of that class role and also to caveats concerning the impartiality of the law as regards non-owners of capital. GMAG is experts' ground too, a place where those whose careers hinge on rapid development in genetic engineering work can lobby for de-restriction. An official safety committee is on management's ground (put there by free booze, low-profile personnel management style and time off as a perk). Watchdog institutions like these are places where we go to keep an eye on others who have effective power which we do not share. Primarily and necessarily, they are sites of refusal.

That boundary is always negotiable, and though negotiation doesn't only take place in formal committees – it also takes place on the ground, in the factory, lab and community – watchdog bodies can be a significant site for the entrenchment and challenging of hegemonies. Sir Gordon Wolstenholme, first chairman of GMAG, knows this, 'Union representatives sometimes seek to judge the scientific merit of proposals for research. The research councils and grant-giving foundations have a difficult enough job to do this by peer review. Neither safety committees nor GMAG are constituted to undertake any such task.' By 'peer review' he means experts policing other experts, the distribution of patronage. There is no good reason why working people should ever have accepted professionals' promises to be good and keep their own house in order, and there is no reason why we should continue to extend that privilege, especially since 'their' house is quite manifestly where we live and work. The issue is not 'scientific merit', but the place of 'scientific' values and experts' privilege in a democratic society.

Within a framework of people's research, watchdog bodies are basically places where we go to watch them, to see what is moving before it is too late to develop our own alternatives through our own institutions of people's research and action. This is a major and general point, because there is a danger that as people become more conscious of the political force of science they will extend to it notions of democracy which are already current in other spheres of culture, as in 'industrial democracy'. Democracy is not having a representative on the company board, or on the local Police Committee, or on the Science and Engineering Research Council, or for that matter, in Parliament. The issue is not how to participate in decisions on 'the good of the business' or 'professional policing' or 'scientific merit' or 'national interests', but how to challenge the specific and narrow interests which define these as separable and 'natural' frameworks and whose hegemony keeps them that way. Who feeds and walks the watchdogs? The state? Then commitment to the watchdog body is commitment, however marginal, to an existing massive apparatus of social control and expert interests. Who watches the watchdogs? Members of Parliament? They may be radical, but they are professionals and stuck with the contradictory and marginal role of radical professionalism. The Unions? Career capital and patronage and detached partisanship all work in the sphere of professional trade unionism too. No established institution is a safe house.

Science is a symbol and a locus of this deep entrenchment of elite power, and people's research is a symbol and an emergent institution of the cultural-political movements that are demanded by the Long Revolution as a democratic-cultural-industrial revolution. 'Non-scientific' is a bogy word. We have to break its hold. Research in biotechnology, for example, might be stopped or limited through political intervention and this would be legitimate. It would be society acting on part of society. The real question of legitimacy is not one of abstract values ('scientific' versus 'non-scientific') but one of struggles around the concrete interests of capital accumulation, state administration, community health, employment and employability, skills and prospects; feeling well, eating well, feeling free, knowing what to do and being free to do it. There already is political intervention in science. Private ownership determines decisions about research in private and public industry. Elite interests under the banner of 'academic freedom' determine 'progress' in academic research. National elites determine priorities in government funding of research. Whether we are dealing with the overtly and heavily political intervention of Lysenkoism, the overt and heavy but 'non-political' intervention of multinationals in the Green Revolution, or the day-to-day intervention of expert and capitalist interests in R&D, the issue is not whether politics has any place in science. It is whose: who says what will be progress?

NOT MORE OF THIS SCIENCE, but different. We have no way of knowing yet what the differences might be in a detailed sense, though we can define them in general terms. They concern market value versus use value, capital accumulation as the dominant criterion of work, the relationships between work and home, metropolis and periphery, thought and action, experts and working people, intellect and violence. We can begin to voice the concerns in debates about alternative products, liberated time, class geography, preconceptualisation and violent science. We can begin to re-organise our cultural resources – our collective power to know and find out – across different sites in class geography mapped in terms of refusal, watchdog science, radical professionalism, people's research. Through that cultural recomposition we may begin to counter and redirect the radical recomposition of science which will take place – in whatever way – during our lifetimes.

In 1954 a famous, gentle man at the centre of the ethical crisis of his time said this:

> If I were a young man again and had to decide how to make a living, I would not try to become a scientist or scholar or teacher. I would rather choose to be a plumber or a peddler, in the hope of finding that modest degree of independence still available under present circumstances.

Albert Einstein – it was he – was a socialist, who knew that 'a planned economy is not yet socialism', that there were cultural relations to be resolved as well as technical ones. In 1954 the Festival of Britain was already beginning to shed its glitter over the homes of working-class Britain. In places now they are almost buried in it – things have come on a lot since then. As Einstein was grappling with his responsibilities as a central figure in a rapidly transforming scientific establishment, capital was gearing up for a massive offensive on working-class culture through both production and consumption, new factories and new goods, and science was at the centre of it in all the ways that this book has explored.

But science will not drag us through this time. The frontiers of exploitation are now far too close to where we live, in our deepest selves. The cutting edge of capital's long revolution is right inside our ideal homes, our established skills, our patterns of life and identities as workers, men and women. The front line of the industrial-democratic-cultural revolution is moving homewards, as definitions of work and home, men's and women's work, education and training, public and private responsibilties, are all renegotiated and relations between the State, private capital and private individuals become more strained and visibly contradictory. A hundred years ago imperialism was the way 'out' for ruling elites. On a global level that strategy continues; but within 'developed' countries the colonisation of private life is also a frontier, where science and society visibly meet. They will be confronted there, or ultimately nowhere, for everywhere else tomorrow is fast being cancelled in the name of progress.

Science or Society has two main arguments which relate to our ability to meet this challenge. First it has argued throughout that science has to be seen as *work*, and experts as workers. This places them right inside all the debates that are now moving around the future of work. The use value of work in relation to 'leisure', education, and working people's whole sense of purposefulness and worth, is becoming a central cultural and political issue. Science is not outside of any of these considerations. The way we want our lives and work to be should determine how we want our sciences to be, not in the abstract, but as specific activities in which resources and efforts are committed to a specific purpose: jobs, careers, values, knowledges.

The second overall argument is that even when it is hard to see what experts actually do at work, and therefore hard to propose alternatives, there can be no difficulty in seeing what they *make*. The products of science, technique and medicine pervade our jobs, our lives and our bodies. When so much power to fix patterns of living, thinking and feeling lies in a single massive institution – 'Science', as the totality of sciences, technologies and medical practices – then elite control of that institution is a danger that democrats cannot continue to live with once they recognise it. Science is not above any battle. Sciences are clearly in the thick of many of the struggles in which we are already engaged – as learners, well or ill people, employed or unemployed workers, householders and home-makers; people trying to live well. There are many special things about sciences which make them often willfully difficult for people outside the specialised work to understand. But this does not mean that sciences can justly claim special privileges as a sector of pro-duction and of culture. If something is valued then we can also revalue it, and if certain autonomies are appropriated by elites then they can be reappropriated in the formation of a common culture.

We need to be clear about sciences' involvement in the struggles we perceive and in the things about modern life which make us feel bad. This book puts forward some ways of looking at scientific work and sciences' products which are meant to help. We need also to be clear about what might be done to make life better, as far as know-ledges are concerned, and who we might work with in this, and how. This book surveys some options and makes some proposals. The point of producing a book which does these things is to promote the sense in readers that science, in practice, urgently needs to be brought within an expanded culture and expanded practice of work-ing people. Sciences are so central in modern life that we must place them where we need them – in our own hands.

High Street Science

SEEWRIGHT OPTICIANS LIMITED
27 High Street   Nontown   PQ7 8TL

EXAMINATION ROOM

WORK SPACE

RECEPTION AREA

WINDOW DISPLAY

THE PRICE OF HIGH STREET SCIENCE

Rent £10,000 – Rates £3,000 – Display signs £1,800 – Fixtures and Fittings £12,000

Once you could buy a pair of specs in Woolworths. But not any more. Opticians have a close monopoly, occupying high-rent High Street sites, in among the building societies and TV/video rental showrooms. (Dollond — his name passed on through the Dollond and Aitchison chain of storefront operations — made an early start with patent rights on achromatic lenses in the late eighteenth century.)

In 1982 a microprocessor controlled lens grinding machine has been marketed in the UK to undercut the entrenched position of skilled optical technicians. (Within Europe such workers are a British peculiarity.) The new system may be operated after half a day's training and operates with an accuracy an order of magnitude greater than conventional manual methods. Lens grinding work thus assumes the characteristic late 20th century form of materials handling plus keyboard punching, plus a worker to schedule and supervise the operators' work.

The price of contact lenses might be cut by half using this technology. But opticians' monopoly of sales probably means that little change in prices will percolate to the High Street.

258

Users' Guide: How to Get Outside This Book

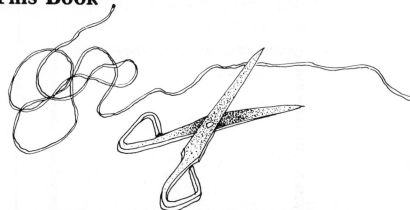

WHERE NEXT? To answer that question you need to know something of where this book is coming from. Immediately it comes from the Channel Four TV series, *Crucible: Science in Society*, but both that series and the ideas and arguments of the book come out of a wider movement of thinking and organising in and around sciences, technologies and medical practices over more than a decade – it sometimes calls itself the radical science movement. This book, therefore, didn't start with words on a page, and it's not our intention – as writer, designer, editor, or publisher – that it should end that way either.

This Users' Guide, therefore, tries to be more explicit than most books are concerning the use of what lies between its covers. We have some suggestions under five headings:

*HELP! What to do if you have problems with the words themselves

*TOPICS How to pursue particular topics further, through books

*RE-READING How to get the most out of the particular organisation of material in this book

*CONTACTS What to do if the issues are new to you but interesting, and you want to discuss them further

*ACTION Where next, outside of reading and talking?

HELP!

Because the issues are complicated and the underlying connections in practice are extensive many of the problems of science in society defy understanding in ordinary terms. At some stage a certain amount of specialist language has to be developed, to cut through commonsense partial understandings. This book therefore has some

259

ordinary words used in special ways consistent with some of their normal usages. This need not interfere with readers' reading. 'Fetish' and 'market value' are words like this: they actually stand for ideas in the theory of ideology and the theory of commodity production, which takes them beyond their everyday senses. The book also has some new or unusual terms which clearly have a specialist meaning and may be unclear if you happen upon them in a particular chapter. 'Preconceptualisation', 'class geography' and 'violent science' are words like this: they are used mainly in one or two particular chapters where their meaning is worked out fairly fully, but they also crop up in others.

To deal with any terms like these, first, use the Index. It has been prepared to serve a double function, as both index and a sort of glossary. If you look up, say, 'culture' you will find page entries for the places where that word is used with a specialist emphasis, so that by scanning them together you can derive your own working understanding of the term. But also there is a list of other related terms: use value, socialisation, values, apparatus. By following these as well, the working definition can be extended. This approach has been adopted rather than a straightforward glossary because all of the words that are really interesting are also really complicated and would demand a small essay to 'define' them. We feel that it's the use of the words which matters, the use value for a particular reader rather than an abstract definition. ('Use value' is in the Index.)

One of the things that radical readers have re-learnt over the last decade and more is that books yield up their use value much better if they are read collectively, in reading groups of any size from two members upwards. Set up for a particular purpose, whether just to read a particular book, or to pursue a particular intellectual or practical issue by tracking down the books which relate to it, reading groups turn the dead words back into living understanding much more effectively than single-person reading usually does. You might find that a reading group makes this book serve you better.

TOPICS
Although they are of very different kinds, each chapter or major sub-section is constructed around a particular topic: technical fixes in agriculture, violence in and around hard science, the work of scientists, new technology in printing, consumerism. Some chapters end with a question: Where is the place of love and life in a monetarist economy? And what do experts do to keep them there? Do we need science to be like this? It is our hope that you will ask many more questions which go outside what the page says, and to help in pursuing the issues (though only you can make the answers) there is a list of possible further reading. The list is not intended to be academically complete, and contains mainly references which come nearest to the interests of the present book. If you should

want to pursue the questions with 'academic' rigour, some of the books will lead you to an academic literature. But don't get lost in it: a reading group again helps to keep reading in contact with purposes.

RE-READING

There is more to the organisation of this book than meets the eye at first reading, and while that is true of any book there is one special sense that applies here. In the radical science movement there are many different groups (some of their addresses are listed in the Addresses section of this book) with many different approaches in theory and in practice. This book has been designed around what seem to be the four main intellectual approaches or themes. These are: sciences as practices of commodity production, sciences as labour processes, sciences as ideological practices, and sciences as practices rooted in and affecting personal identity. More will be said about each of these in a moment. Most of the chapters were designed to sit mainly within one or other of these themes – see the 'Map' of the book overleaf. This has meant that any particular topic has received less than its full treatment, because real issues do not sit nicely within intellectual boundaries. The themes haven't been slavishly pursued, because the issues do matter more, but there is sufficient difference between chapters for you to be able to get some of the feel of each characteristic theme. Each has a cluster of key ideas.

Theme 1 is *capital*: market value, profit, commodities, economic exploitation, capital accumulation, and the reproduction of a social order dominated by private ownership and profit. This is the characteristic approach of much of the radical work and many campaigns around Agribusiness, for example, on work hazards, or environmental pollution, or military science. As a theme it cannot stand properly alone, but the same is true of all the others. The streams of thinking need to be worked together, in practice, in order to build effective understandings and oppositional practices.

Theme 2 is *work*: skills, the socialisation of production, hierarchy, fragmentation, routinisation, the connections between practices through apparatus. These are key issues in what radicals over the past five years or so have been calling a 'labour process' approach. The watchword of these approaches is often 'deskilling', but in the context of this book the attempt has been made to put this sometimes narrow interest into a broader relationship with the other themes. Capital investment, for example (Theme 1) is the main determining factor in the apparatus of work. Theme 2 is vital because the apparatus of work, home and school life imposes very severe limits on what alternatives may be effectively struggled for at any time; but a necessary complement is the third theme.

Theme 3 is *class*: the interests and the power of classes and elite

261

☐ A MAP OF THE BOOK

1 commodities /capital **2** labour processes **3** ideologies/ hegemony **4** identity

○ Tomorrow has been cancelled

① Mankind's little helper?

② Science & the People (1): Science 'On the Grain Front'

③ Science & the People (2): 'The Green Revolution'

④ Scientific workers

⑤ Uses of science

⑥ Tools, money & careers

○ Managers must measure

⑦ Schooling

⑧ A whole history of inventions

⑨ Design of jobs

⑩ Violent science

⑪ Consumers & knowledge

○ Progress!

⑫ Living well: democracy and science

groups, use values, struggles in and around science, technology and medicine, the entrenchment of interests in practices. One key emphasis in this line of thinking is on the power of culture and the way that this links with the existence of 'experts' in most spheres of social life. Another key emphasis often encountered is 'ideology' – the use of scientific work as a basis in theory and practice for advancing specific classes, elite or sectoral interests. The aim of this book has been to absorb ideology, as a focus, within the notion of hegemony, and thus to link struggles in ideas with struggles in apparatus: this is the link between Theme 3 and Theme 2. As an approach to questions of class and struggle, Theme 3 also has obvious though complex links with Theme 1, capital.

Theme 4 is, in contemporary culture and in the radical science movement, the least obvious. It is *identity*: feelings of powerlessness, inferiority, anxiety, issues of emotionality, gender, race, oppression and repression, 'subjectivity' (and therefore also 'objectivity'). This is a focus powerfully introduced into discussions and struggles of science by the women's movement, still tenuously related to more established themes and particularly under-theorised – hardly surprisingly, given its complexity and the forces to which it relates. This emphasis is vital to struggles for alternatives around science, because so much depends on people's sense of what they can know, their own personal sense of potency. The arguments link, clearly, with issues about experts and working people (Theme 3). They link, as in the discussion of schooling, with arguments about the division of labour (Theme 2). They link, as in the discussion of violent science, with arguments about the power of capital (Theme 1). One main ambition of this book is to bring these lines of argument and areas of struggle closer together.

You might use this four-theme organisation of the book in two ways. First, in re-reading, you might choose to read thematically, say, within Theme 2 – work and the division of labour. In this way the strengths, limitations and connections of that theme might become clearer. You could do the same with the other themes. Because there is a variety of content within each theme these divisions might be used as a basis for short programmes of private or collective, formal or informal study at home, in reading groups, study groups, adult education classes, trade union courses, women's groups, schools, and so on.

Secondly, again in re-reading, you might follow topics across themes using the Index, and in this way see how particular aspects of each topic show up under the different approaches of the four themes. You might follow 'narrowness in knowledge', for example, and find it in all four streams of discussion. You might follow 'consumers' or 'experts' or 'jigsaw puzzling'. The hope is that you will be able to assess the strengths and weaknesses of the different approaches, the better to be able to use them in concert and to go beyond them.

CONTACTS

The Science in Society unit of Central Television, which produces the *Crucible* series, was set up with the intention of generating more than one-off TV programmes. Books – like this one – were a second line of work, and videocassettes will be another. But in addition to all these particular products the Unit intends to initiate and advance debates, to cultivate public awareness, and to help to develop popular participation in evaluating sciences' social priorities and directing scientific, technological or medical developments. There are three direct and simple ways that this purpose might be served in your use of this book.

First, if its approaches and topics catch your interest we would like to know what exactly interests you most, so that further work can be more usefully directed. The contact address is:

P.O. Box 280, London N7 9RX

Second, if your interest is a new one, or if you live in a place where there seem to be no groups or organisations concerned with issues of the kind with which this book deals, we may be able to help you get in touch with other people with similar interests. If you write to the Crucible unit asking to be put in touch – or offering to be contacted – we can prepare a directory of names and addresses in this way, and mail it to those whose names are in it. This is a basis – though only a bare basis – for a learning exchange. Initiatives on that basis are necessarily up to you, though it is conceivable that the Unit might be able to help further in other ways.

Third, if this book goes to a second edition it should, of course, become a better and more useful thing. It can only really do this if you let us know how the present book serves your interests and purposes. Write and tell us, or send us a cassette letter.

The general aim of all these suggestions is to turn the dead products – the book and the TV programmes – back into living understandings and actions. There are more ambitious versions of this aim too, which may be of interest to you. Unlike TV programmes screened over the air, videocassettes are a good basis for discussion: they can be specially designed to talk around, perhaps with an accompanying speaker, and they can be stopped, replayed and reused as necessary at the demand of the users. You may want to use the *Crucible* videocassettes by forming a neighbourhood video club, or in an evening class, or in a trade union branch meeting, to prompt discussion of issues connected with science or technology or medicine. Information on videocassettes is available from the address given above. If groups want to have a speaker to address one of their meetings then the Unit may be able to help with this too, by putting them in touch with the Science in Society Education

Co-ordinating Committee, formed for this purpose.

ACTION
It would contradict the whole tone of this book if it were to end with an attempt to prescribe for you, the unknown reader, what course of action might be appropriate where you happen to live and work. There have been examples and suggestions – particularly in chapter 12 – and some of them are summed up in the 'Places and Voices' chart at the end of this Guide, which is intended only to prompt and support thinking, not to define approaches and issues.

Action implies organisation, and between them the radical science movement, the women's movement and the labour movement have many formal organisations as well as many forms of organisation. The book has discussed some of the forms of organisation as they relate to knowledge (they are summed up in the 'Places' headings on the chart) but to discuss the formal organisations in any depth would be impossible because of their number and geographical spread. Therefore we have had to make do with including a selected list of addresses which you may be less likely to know of. The list follows the Further Reading. We have not included political parties, major national campaigns (such as CND), academic institutions, official trade union offices and government departments and research institutes, because we think that it is relatively easy to find out about them through personal contacts, local libraries, advice centres, and so on.

There is one organisation to which we want to draw special attention, because it illustrates more fully than anything else in Britain what developed people's research may look like. This is the *Network of Labour and Community Research and Resource Centres* (see the Addresses section). Set up in 1980 it links centres in eight English cities, and is steadily being expanded. The centres offer help with research and investigation, specialist advice on money, jobs, law, health and safety, and other matters, help in organising campaigns, information services and briefings, and 'publishing' assistance to groups of many kinds: tenants' groups, women's organisations, ethnic groups, shop stewards' committees, unemployed workers' groups, trade union branches, trades councils, regional union committees and TUCs. Their focus in practice is in some ways broader than the focus of this book and in some ways narrower: broader in areas of law, local government, employment issues and so on, but narrower in their necessary concentration on the visible aspects of technologies rather than the associated knowledges. We recommend the Network to you, as a source of help if it is within your reach and as one model of organisation whether its centres are near you or not. A prospectus is available from the contact address.

The importance of the centres in the Network lies in the way they work. As they describe it:

265

☐We are accountable to the groups with whom we work, and work in close dialogue with them

☐We do not come in as 'outside experts' but work alongside the groups as committed advisers

☐We try to build up the strength and self-sufficiency of the groups we work with by passing on our knowledge, skills and experience in as understandable and useful ways as possible.

These are touchstones of people's research, and constitute a checklist of points to watch whenever experts are involved in a practical issue.

The book has tried to make it clear that there are, historically, any number of thinkable alternatives to what we know as sciences and their ways of doing things. But alternative and oppositional values must be firmly entrenched in organisation if they are to make any real difference. 'Palace revolutions' will not change science. Nor will revolutions *for* the people, in or out of science. It may seem odd to talk of moving on, even before critical thinking about science has succeeded in becoming a major emphasis in our culture. Nevertheless, move on we must, beyond science, to *work*, so that what we criticise and act through is a significant whole rather than a significant part. We need to look beyond ideas and ideology to practices and hegemony, so that we do not content ourselves with superficial debates and changes. We need to break with the idea of knowledge as the kind of thing which can be lumped into a 'body' – or indeed, as any kind of thing at all, so that we can develop a powerful, personal and intimate understanding of knowing, as a collective activity which goes wider than work and deeper than consumption. Capitalist work is not the making of a future, just the reproduction of an eternal present. Capitalist work – and therefore capitalist science – can only ever be more or less but never any better.

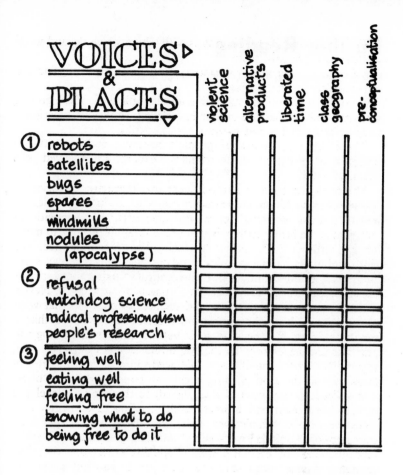

	violent science	alternative products	liberated time	class geography	pre-conceptualisation
① robots					
satellites					
bugs					
spares					
windmills					
nodules					
(apocalypse)					
② refusal					
watchdog science					
radical professionalism					
people's research					
③ feeling well					
eating well					
feeling free					
knowing what to do					
being free to do it					

USING THE GRID
This is just to prompt exploration of connections.

☐ Move from the places where the issue seems to be

☐ across to the voices which seem to be appropriate
in exploring the issue

☐ then up and down to other places where the issue
needs to be located for debate or action. Repeat.

① is 'their' ground   ② is the ground of
③ is the people's ground   struggles/fragments

See Index for terms used in the grid.

Further Reading

Unless otherwise stated, books are published in London

In addition to a chapter by chapter listing there is a general section at the end. This is not a complete list but merely contains some suggestions, most of them close to the point of view of this book.

TOMORROW HAS BEEN CANCELLED

British Society for Social Responsibility in Science, *Nuclear Power:The Rigged Debate*, BSSRS, 1980. From BSSRS, 9 Poland Street, London WC1V 3DG, 60 pence + p&p. Criticises the establishment and also some of the conventional ways of criticising.

Conference of Socialist Economists, *Microelectronics: Capitalist Technology and the Working Class*, CSE Books, 1980. Centres on 'robots' rather than 'satellites'.

Dave Elliott and others, *The Politics of Nuclear Power*, Pluto Press, 1978. Deals with 'windmills' from the point of view of employment and industrial strategy.

Watson Fuller, ed., *The Social Impact of Modern Biology*, RKP, 1971.

Green Alliance (Richard Boston, Richard Holme and Richard North, eds), *The Little Green Book: An Owner's Manual to the Planet*, Wildwood House, 1979. From Green Alliance, 60 Chandos Place, London WC2N 4HG, 95p retail (50p for ten or more copies). A handy reference book, especially on 'windmills' and 'nodules'.

Edward Yoxen, 'Life as a Productive Force: Capitalising the Science and Technology of Molecular Biology', in Levidow and Young, vol. 1 (see general reading below), 67–122. A history of 'bugs'.

1 MANKIND'S LITTLE HELPER?

Jasia Reichardt, *Robots: Fact, Fiction and Prediction*, Thames and Hudson, 1978.

Studs Terkel, *Working*, Penguin, 1977. 'People talk about what they do all day and how they feel about what they do.'

Raymond Williams, *The Long Revolution*, Harmondsworth, Penguin, 1965. A classic British probing of ideas in the making of a common culture.

2 SCIENCE 'ON THE GRAIN FRONT'

Kendall E. Bailes, *Technology and Society Under Lenin and Stalin: Origins of the Soviet Technical Intellegentsia 1917–41*, Princeton University Press, Princeton, NJ, 1978.

Nicholas Lampert, *The Technical Intellegentsia and the Soviet State: A Study of Soviet Managers and Technicians 1928–35*, Macmillan, 1979.

Richard Lewontin and Richard Levins, 'The Problem of Lysenkoism', in Hilary Rose and Steven Rose, ed., *The Radicalisation of Science*, Macmillan, 1976, 32–64.

Bob Young, 'Getting Started on Lysenkoism', *Radical Science Journal* 6/7 (1978), 81–105. A guide to literature on the USSR and approaches to Lysenkoism in the history of science.

3 'THE GREEN REVOLUTION'

Franz Fanon, *The Wretched of the Earth*, Harmondsworth, Penguin, 1978.

Teresa Hayter, *The Creation of World Poverty*, Pluto Press, 1981.

Susan George, *How the Other Half Dies*, Harmondsworth, Penguin, 1976.

Michael Kidron and Ronald Segal, *The State of the World Atlas*, Pan/Pluto Press, 1981.

Andrew Pearse, *Seeds of Plenty, Seeds of Want: Social and Economic Implications of the Green Revolution*, Oxford, Clarendon Press, 1980.

5 USES OF SCIENCE

John Desmond Bernal, *Science in History*, 4 vols, illus., Harmondsworth, Penguin, 1969.

Ruth Schwartz Cowan, 'More Work for Mother: The Labour Process and Technology in the American Home', in Levidow and Young, vol. 2 (see general reading below).

Brian Simon, *The Two Nations and the Educational Structure 1780–1870*, Lawrence and Wishart, 1974.

MANAGERS MUST MEASURE

John Irvine, Ian Miles and Jeff Evans, eds, *Demystifying Social Statistics*, Pluto Press, 1979. A powerful collection of articles on aspects of the numbers game, from the Scientific Revolution to cost-benefit analysis, including accounts of official statistics and the use of statistics in counter information.

Radical Statistics Nuclear Disarmament Group, *The Nuclear Numbers Game: Understanding the Statistics Behind the Bombs*, Radical Statistics, 1982. From BSSRS, 9 Poland Street, London W1V 3DG, £1.50 + 35p.

7 SCHOOLING

Graz Baran, 'Why Care About Physics?', *Teaching London Kids* 18, from 40 Hamilton Road, London SW19, 50p p&p.

Douglas Barnes, *From Communication to Curriculum*, Harmondsworth, Penguin, 1976.

Ivan Illich, *Tools for Conviviality*, Fontana, 1975.

Peter Medway, *Finding a Language: Autonomy and Learning in School*, Writers and Readers/Chameleon, 1980.

Wilhelm Reich, *Listen, Little Man!*, Harmondsworth, Penguin, 1975.

Lord Todd, O. M., Presidential Address, 30 November 1978, *Proceedings of the Royal Society of London*, A, 365, v–xviii.

Paul Willis, *Learning to Labour: How Working Class Kids Get Working Class Jobs*, Saxon House, 1977.

8 A WHOLE HISTORY OF INVENTIONS

Maxine Berg, ed., *Technology and Toil in Nineteenth Century Britain*, CSE Books, 1979, A collection of documents.

Cynthia Cockburn, *Brothers: Technology and Trade Unionism in Printing – A Socialist Feminist Account*, Pluto Press, 1982, forthcoming.

Campaign Against Depo Provera, *Report*, available c/o ICAS, 374 Grays Inn Road, London WC1.

David Dickson, *Alternative Technology and the Politics of Technical Change*, Fontana, 1974.

Mike Duncan, 'Microelectronics: Five Areas of Subordination', in Levidow and Young, vol. 1 (see general reading below), 172–207.

Siegfried Giedion, *Mechanization Takes Command*, New York, Norton Library, 1969.

Dick Hebdige, *Subculture: The Meaning of Style*, Methuen, 1979.

Hari John, 'Medical Imperialism, Feminism and the Third World: An Interview', *Spare Rib* 116 (March 1982), 49–51.

Karl Marx, *Capital*, vol. 1 (especially chs. 13–15), Harmondsworth, Penguin/New Left Review, 1976.

Pacific Research Center, *Delicate Bonds: The Global Semiconductor Industry*, Mountain View, California, 1981. A special issue of *Pacific Research*, from 867 West Dana Street, No. 204, Mountain View, CA 94041, USA: $1.50 ($2.00 foreign air mail).

Pacific Research Center, *The Changing Role of S. E. Asian Women: The Global Assembly Line and the Social Manipulation of Women on the Job*, special joint issue of *Pacific Research/S.E. Asia Chronicle*, 1978/79. From above address, $1.00.

Jill Rakusen, 'Depo Provera: A Case Study in the Politics of Birth Control', in Helen Roberts, ed., *Women, Health and Reproduction*, 75–108.

Edward Thompson, 'Time, Work-Discipline and Industrial Capitalism', *Past and Present* 38 (1967), 56–97.

Edward Thompson, *The Making of the English Working Class*, Harmondsworth, Penguin, 1968.

Raymond Williams, *The Country and the City*, Paladin, 1975. About the invention of golden ages, landscape, rural idylls, urban masses.

9 DESIGN OF JOBS

Huw Beynon, *Working for Ford*, Harmondsworth, Penguin, 1973.

Mike Cooley, *Architect or Bee: The Human-Technology Relationship*, Hand and Brain Publications, 1980. From Langley Technical Services, 95 Sussex Place, Slough, Surrey SL1 1NN.

Andre Gorz, ed. *The Division of Labour: The Labour Process and Class Struggle in Modern Capitalism*, Hassocks, Harvester Press, 1976.

John Heskett, *Industrial Design*, Thames and Hudson, 1980.

Philip Kraft, *Programmers and Managers: The Routinisation of Computer Programming in the United States*, New York, Springer-Verlag, 1977. ISBN 0 387 90248 1.

Theo Nichols, ed., *Capital and Labour: Studies in the Capitalist Labour Process*, Fontana, 1980.

Harry Braverman, *Labor and Monopoly Capital: The Degradation of Work in the Twentieth Century*, Monthly Review, 1974.

10 VIOLENT SCIENCE

Carol Ackroyd and others, *The Technology of Political Control*, 2nd edn., Pluto Press, 1980.

Stephan Chorover, 'The Pacification of the Brain: From Phrenology to Psychosurgery', in T. P. Morley, ed., *Current Controversies in Neurosurgery*, W. B. Saunders Co., 1976, 730–67. Shows how scientists have taken the US Army officer's Vietnam apology, 'It was necessary to destroy the village in order to save it', as a text for brain research.

Brian Easlea, *Science and Sexual Oppression*, Weidenfeld and Nicholson, 1981.

Mary Kaldor, *The Baroque Arsenal*, Deutsch, 1982.

William Leiss, *The Domination of Nature*, New York, Braziller, 1972.

Carolyn Merchant, *The Death of Nature: Women, Ecology and the Scientific Revolution*, San Francisco, Harper and Row, 1980.

William Carlos Williams, *In the American Grain*, Penguin, 1972.

Tom Wolfe, *The Right Stuff*, New York, Bantam, 1980.

Solly Zuckerman, *Nuclear Illusion and Reality*, Collins, 1982.

11 CONSUMERS AND KNOWLEDGE

BSSRS Agricapital Group, *Our Daily Bread: Who Makes the Dough*, BSSRS, 1978. Available from BSSRS, 9 Poland Street, London W1V 3DG, 50p (institutions £5.00).

N. I. Bukharin and others, *Science at the Crossroads*, Papers from the Second International Congress of the History of Science and Technology, 1931, by delegates of the USSR, reprinted by Frank Cass and Co., 1971. See also special feature in *Science for People* 51 (Spring 1982), papers celebrating the 50th anniversary of the 1931 meeting; from BSSRS, 9 Poland Street, London W1V 3DG, 70p + p&p.

Barbara Ehrenreich and John Ehrenreich, 'The Professional-Managerial Class', in Pat Walker, ed., *Between Capital and Labour*, Hassocks, Harvester Press, 1979, 5–45.

Ivan Illich, *The Right to Useful Unemployment and Its Professional Enemies*, Marion Boyars, 1978.

Radical Science Journal Collective, 'Science, Technology, Medicine and the Socialist Movement', *Radical Science Journal* 11 (1981), 3–70. Discusses fifty years of radical British approaches to science.

Gary Werskey, *The Visible College*, Harmondsworth, Allen Lane, 1978. A collective biography of five 1930s radical scientists.

12 LIVING WELL: DEMOCRACY AND SCIENCE

Reyner Banham, *Design by Choice,* Academy Editions, 1981.

Boston Women and Health Book Collective, *Our Bodies, Ourselves*, Harmondsworth, Penguin, 1980.

Boston Women and Health Book Collective, *Ourselves and Our Children*, Harmondsworth, Penguin, 1981.

Charlie Clutterbuck and Tim Lang, *More Than We Can Chew: The Crazy World of Food and Farming*, Pluto Press, 1982, forthcoming.

Cynthia Cockburn, *The Local State*, Pluto Press, 1978.

Lesley Doyal with Imogen Pennell, *The Political Economy of Health,* Pluto Press, 1979.

Mike Hales, *Living Thinkwork*: *Where do Labour Processes come from?* CSE Books, 1980.

IWC Motors Group, *A Workers' Enquiry Into the Motor Industry*, CSE Books, 1978.

Lancashire School Meals Campaign, *Now You See Them, Now You Don't*, A Report on the Fate of School Meals and the Loss of 300,000 jobs, available from LSMC Secretary, 17 Marlowe Avenue, Baxenden, Accrington, Lancashire, England, £1 (institutions £2).

Les Levidow, 'Grunwick: The Social Contract Meets the Twentieth Century Sweatshop', in Levidow and Young, vol. 1 (see general reading below), 123–171.

London-Edinburgh Weekend Return Group, *In and Against the State: Discussion Notes for Socialists*. Distributed by Southern Distribution/ Scottish and Northern Books, ISBN 0 9506769 O X.

Victor Papanek, *Design for the Real World*, Paladin, 1974.

Trades Councils, *State Intervention in Industry: A Workers' Inquiry*, Spokesman Books, second edition, 1982, forthcoming.

Hilary Wainright and Dave Elliott, *The Lucas Plan: A New Trade Unionism in the Making?* Allison and Busby, 1982, forthcoming.

GENERAL

David Albury and Joseph Schwartz, *Partial Progress: The Politics of Science and Technology*, Pluto Press, 1982, forthcoming.

Rita Arditti, Pat Brennan and Steve Cavrak, eds., *Science and Liberation*, Boston, South End Press, 1980.

BSSRS London Group, *Science On Our Side*, BSSRS, 1982, forthcoming.

BSSRS Todmorden Group, *Facing Up to Science*, BSSRS, 1982, forthcoming. Both available from BSSRS, 9 Poland Street, London W1V 3DG.

Les Levidow and Bob Young, eds., *Science, Technology and the Labour Process: Marxist Studies*, 2 vols. CSE Books/Humanities Press, 1981 (vol. 1), 1982, forthcoming (vol. 2). Volume 2 contains articles on the construction of social reality through technology, two essay reviews of Braverman's *Labor and Monopoly Capital*, a reappraisal of Luddism in relation to New Technology, an exploration of the limits of labour process thinking, a historical review of the recomposition of housework/wage work in America, Italian analyses of 'windmills' technology as capitalist recomposition, and an account of the class geography of two diseases, one in nineteenth-century women, one in coal miners.

C. H. Waddington, *Tools for Thought*, Paladin, 1977. An anti-reductionist biologist surveys 'systems' ways of looking at the modern world. Says nothing about science in society, but even in their abstractness the ideas are important ones to come to terms with.

Raymond Williams, *Keywords*, Fontana, 1976. A hundred or so major modern words and how they got here: collective, class, democracy, hegemony, improve, modern, nature, progressive, radical, science, tradition, work, are some of them.

Some Addresses

This list excludes political parties, major national campaigns such as CND, academic institutions or adult classes, official trade union offices, research institutes and government departments. A local library, advice centre or people's bookshop will be able to help on these. Listed here are some of the organisations most closely allied with the kind of politics implicit in this book but least widely known.

British Society for Social Responsibility in Science (BSSRS), 9 Poland Street, London W1V 3DG. Apart from publications by working groups, some of which are included in the Further Reading list, BSSRS produces *Science for People*, a magazine with fifty issues over ten years. BSSRS has topic groups and local groups. Local groups exist in: Birmingham, Brighton, Cambridge, Durham, Edinburgh, Kingston, London, Manchester, Sheffield and a trans-Pennine (Todmorden) group. BSSRS working groups are: Agricapital (publishes *Food and Politics*), Politics of Energy, Sociobiology, Work Hazards (publishes *Hazards Bulletin* bi-monthly and booklets on Noise, Oil Mists, Vibration and Asbestos), Hospital Hazards, Radiation Hazards, Women and Work Hazards, Women's Caucus, Unemployment among Scientists, Science Teachers. Affiliated groups are: Politics of Health Group, (publishes a newsletter, £2 + six large stamped self-addressed envelopes), Radical Statistics Group, Radical Statistics Health Group and Radical Science Journal. Membership, including *Science for People* and members' internal bulletin, is £7.00 yearly (£3.50, students and unwaged people): non scientists and technical workers are not excluded!

Radical Science Journal, 26 Freegrove Road, London. Aims to provide a forum for extended analyses of the ideology and practice of science, technology and medicine from a radical political perspective. Most contributors have attempted to re-examine past marxist views and develop a marxist critique of scientism in the socialist movement. In addition to producing *Radical Science Journal* the collective organises series of seminars in London. The journal is £6.00 for three issues, institutions £13.50. Add £0.60 to foreign cheques.

Association of Radical Midwives, c/o Sally Hart, Women's Centre, 40 Turnpike Lane, London N.8.

Centre for Alternative Industrial and Technological Systems (CAITS), c/o North East London Polytechnic, Longbridge Road,

Dagenham, Essex RM8 2AS. Set up at the instigation of shop stewards from the Lucas Aerospace combine, the unit is mainly engaged in research on behalf of combine and shop stewards' committees. Publishes a number of leaflets on alternative products, workers' plans, conversion of military production and similar topics.

Unit for Development of Alternative Products, Department of Combined Engineering, Lanchester Polytechnic, Priory Street, Coventry, West Midlands.

Popular Planning for Social Need. A declaration sponsored by five trades councils and four combine committees, calling for a campaign to fuel a debate in the labour movement. Further information from the coordinator, Jane Barker, c/o CAITS (address above).

Union of Physically Impaired Against Segregation (UPIAS), c/o Flat 2, Saint Giles Court, Dane Road, London W13.

Group for Alternative Science and Technology Strategies (GASTS). Meets in the North and West Midlands to discuss science and technology in the context of alternative economic strategies for the labour movement, and to propose options for wider discussion. Further information from the secretary, Ken Green, 9 Dalston Drive, Manchester 20.

Conference of Socialist Economists (CSE), 25 Horsell Road, London N5 1XL. Less academic than its now-outgrown name suggests, this is a federation of left groups and individuals committed to developing a materialist critique of capitalism in marxist terms within the labour movement. Publishes *Capital and Class* quarterly. Working groups on topics including Money, Housing, State Expenditure and the Cuts, the Law, Sex and Class; also local groups from Brighton to Edinburgh. Membership, including *Capital and Class* and newsletter, is £7.00 full, £4.50 reduced, £15.00 supporting (overseas: £8.00, £4.50, £15.00. Add £1.00 for clearing foreign currency cheques. Airmail sub £16.00).

Socialist Environment and Resources Association (SERA), 9 Poland Street, London W1V 3DG. Advances discussion of science and technology issues – notably energy policy and employment – in the labour movement; has consultative status with the Labour Party.

Intermediate Technology Development Group, 9 King Street, Covent Garden, London WC2. Concerned with alternative technologies in non- and less-industrialised countries.

Undercurrents, 27 Clerkenwell Close, London EC1. A magazine focusing on alternative technologies and anti-nuclear issues.

Friends of the Earth, 9 Poland Street, London W1V 3DG. A national organisation with local groups 'using all available means to change current attitudes and policies' concerning the environment. Publishes many leaflets and books via Earth Research Resources Ltd, 40 James Street, London W1.

Network of Alternative Technology and Technology Assessment (NATTA), Alternative Technology Group, Faculty of Technology, Open University, Walton Hall, Milton Keynes, Bucks.

Network of Labour and Community Research and Resource Centres, c/o coordinator, 118 Workshop, 118 Mansfield Road, Nottingham. Centres in Tyneside, Bradford, Leeds, Manchester, Nottingham, Coventry, Bristol and London, and gradually expanding. See comments in Users' Guide above, under 'action'.

Counter Information Services (CIS), 9 Poland Street, London W1V 3DG. Publishes many reports on aspects of capitalism, institutions and companies, and ruling ideas. Provides research and information services.

State Research, 9 Poland Street, London W1V 3DG. Keeps tabs on the State and its repressive agencies and publishes bi-monthly bulletin.

Armament and Disarmament Information Unit, Science Policy Research Unit, University of Sussex, Falmer, Brighton, Sussex.

Social Audit, 9 Poland Street, London W1V 3DG. The publishing arm of Public Interest Research Ltd, which conducts research into government and corporate activities.

Women's Research and Resources Centre, 190 Upper Street, London N1.

Collective Design/Midland, c/o 23 Gordon Street, Leamington Spa, West Midlands CV31 1HR.

Trade Union International Research and Education Group (TUIREG), Ruskin Hall, Duncton Road, Old Headington, Oxford OX3 9BZ.

Federation of Worker Writers and Community Publishers, 'E' Floor, Milburn House, Dean Street, Newcastle-upon-Tyne NE1 1LF. There are many affiliated groups across the country: or why not start one?

Index

For further links see related entries in square brackets.

277